Ethernet in the First Mile

The IEEE 802.3ah EFM Standard

Michael Beck

McGraw-Hill

New York   Chicago   San Francisco   Lisbon   London   Madrid
Mexico City   Milan   New Delhi   San Juan   Seoul
Singapore   Sydney   Toronto

The McGraw·Hill Companies

Library of Congress Cataloging-in-Publication Data

Beck, Michael.
    Ethernet in the first mile : the IEEE 802.3 ah EFM standard / Michael Beck.
      p. cm.
    Includes bibliographical references and index.
    ISBN 0-07-145506-X
    1. Ethernet (Local area network system)   I. Title.

    TK5105.8.E83B425   2005
    004.6'8—dc22

                                   2005045579

1 2 3 4 5 6 7 8 9 0 DOC/DOC 0 1 0 9 8 7 6 5

ISBN 0-07-145506-X

The sponsoring editor for this book was Stephen S. Chapman and the production supervisor was Sherri Souffrance. It was set in Century Schoolbook by Patricia Wallenburg. The art director for the cover was Anthony Landi.

Printed and bound by RR Donnelley.

This book was printed on recycled, acid-free paper containing a minimum of 50% recycled, de-inked fiber.

McGraw-Hill books are available at special quantity discounts to use as premiums and sales promotions, or for use in corporate training programs. For more information, please write to the Director of Special Sales, McGraw-Hill Professional, Two Penn Plaza, New York, NY 10121-2298. Or contact your local bookstore.

Disclaimer and Notices

Most standards and draft standards are copyrighted documents, owned by their Standards Development Organization (SDO). This book references several standards (listed in the references), but makes no claim to describe all the requirements for compliance with these standards. For any compliance issues, the original standard is the only authority. Only the responsible SDO is competent to provide authoritative interpretations of any portion of a standard. To purchase an up-to-date copy of a standard, contact the SDO that developed it (see the References section for details).

This book's main focus is on IEEE Std. 802.3ah-2004, as developed by the "Ethernet in the First Mile" Task Force. It further explains technology specified in other standards or draft standards. Although this book contains useful information for users and implementers of "Ethernet in the First Mile" equipment, it is no substitute for IEEE Std. 802.3 and its amendments. "IEEE" and "802" are registered trademarks of IEEE, Inc. Other terms used in this book may be trademarks.

Some of the algorithms, methods, and apparatuses described in this book may be protected by intellectual property rights, such as U.S. or foreign patents or patent applications.

The views and opinions expressed in this book are those of the author alone, and do not represent the views of the publisher of this book or any corporation with which the author may be associated.

Figure 1.1 reprinted with permission from IEEE Std. 802-2001, "Overview and Architecture," by IEEE. Figure 10.1 reprinted with permission from IEEE Std. 802.1D-2004, "Media Access Control (MAC) Bridges," by IEEE. Figure 10.2 reprinted with permission from IEEE Std. 802.1Q-2003, "Virtual Bridged Local Area Networks," by IEEE. Table 9.1 (adapted) reprinted with permission from IEEE Std. 802.1X-2001, "Port-Based Network Access Control," by IEEE. Table 8.5

Contents

Preface

This book is the result of almost four years of close involvement in the creation of the "Ethernet in the First Mile" (EFM) standard. Working for a company with a strong xDSL focus, I was initially sent to an EFM meeting to monitor the progress in the field of Ethernet-over-xDSL, and to assess the impact on existing xDSL standards projects. Exploration soon turned to participation. With the members of the IEEE 802.3 Working Group as my tutors, I familiarized myself with the underlying concepts of Ethernet, and the IEEE standards development procedures. I became a part of an enthusiastic team of Ethernet and xDSL experts, making their contribution to the global evolution towards packet/frame-based networks. My commitment was formalized when I accepted the task of Editor of the Copper Sub Task Force.

In the process, I often found myself explaining xDSL concepts to people with an Ethernet background, and Ethernet concepts to people with an xDSL background. This led me to the idea of writing a tutorial on Ethernet-over-xDSL, explaining all the necessary Ethernet and xDSL concepts to people with a basic understanding of data communications. Combining this tutorial with material on the optical and OAM specifications, I had the rough outline of a book on EFM.

The process that leads to the creation of a standard is a long and complex one. This book discusses the process in more detail than other books, because I believe it will help the reader understand why the standard is the way it is. Chapter 1 is therefore dedicated to the standardization process and the place of the EFM standard in the Ethernet ecosystem. Notes on the history of specific features of the standard are provided in the other chapters where relevant. As for the standardization of a single short-reach copper solution (10PASS-TS), the Ethernet in the First Mile Task Force has succeeded where other Standards Development Organizations (SDOs) have failed, and it is worth exploring how this happened. If in doing so I can convince some readers to actively participate in standards development in the future, that would be a welcome result.

The rest of the book is organized as follows. Chapter 2 is a specialized chapter on the topic of MAC-PHY rate matching. Although it may not seem to be sufficiently substantive to deserve a separate chapter, it relies on concepts explained in Chapter 1 to describe a feature of EFM used by the copper and fiber PHYs of the later chapters. The copper PHYs are discussed in detail in Chapter 3 (general characteristics), Chapter 4 (the short-reach solution: 10PASS-TS), and Chapter 5 (the long-reach solution: 2BASE-TL). The fiber-based solutions are described in Chapter 6 (general characteristics and point-to-point PHYs) and Chapter 7 (Ethernet Passive Optical Networks). Operations, Administration, and Maintenance (OAM) are discussed in Chapter 8, which also deals with management issues.

For completeness, two topics that are outside the scope of the Ethernet in the First Mile project were added to the book. The security standards discussed in Chapter 9 were directly inspired by the EPON work in the Ethernet in the First Mile Task Force. The bridging standards discussed in Chapter 10 are independent of any specific LAN technology, but they do deserve a place in a book that attempts to give an insight in the way Ethernet is used to build networks in the real world.

Writing a book like this one is a balancing act between providing too many details, thus rewriting the standard, and summarizing too much, thus taking things for granted that may not be obvious to the majority of the readers. Some topics receive more attention because they are fundamental or intertwined with other important issues. Where concepts are assumed to be widely known, less explanation is provided.

As the Ethernet in the First Mile standard is a mosaic of references to existing standards, combined with a good deal of original work, I imagine that this book can be useful both for readers of the standard and for people who want to learn about Ethernet in the First Mile without reading the standard. The detailed bibliography at the end of the book will guide the interested reader to the documents used in the development of the Ethernet in the First Mile standard, and to books and articles explaining specific aspects of Ethernet in more detail. Throughout the book, figures in brackets refer to the items in the references section that are relevant to the topic under discussion. Visit www.ethernetinthefirstmile.com for useful links and the most recent information.

Acknowledgments

I have had the pleasure of working on an Ethernet-over-VDSL prototype at Alcatel Bell during the period in which the foundations of the Ethernet in the First Mile standard were being laid in the IEEE 802.3ah Task Force. This has been a very rewarding experience, thanks to team members Thierry Pollet, Piet Vandaele, Eric Borghs, Jeroen Jacobs (now with Niko), and Adrian Mihanta (now with Siemens VDO). The prototype was successfully demonstrated at the EFM Alliance demo event in San Francisco in July 2003.

I am also grateful to colleagues Tom Van Caenegem, François Fredricx, Tim Gyselings, Edwin Ringoot, and Duane Remein for helping me discover the intricacies of (E)PON, and to Erwin Six and Jeanne De Jaegher (now with D.G. Information Society and Media of the European Commission) for our countless discussions about network architecture issues. Alcatel's xDSL standards experts Sigurd Schelstraete (now with Ikanos), Danny Van Bruyssel, Frank Van der Putten, J. Lane Moss, Frank Ploumen, Peter Reusens (now with LEA), and Jan Verlinden have shared valuable insights in xDSL standardization with me.

I have been able to find my way around IEEE Working Group 802.3, thanks to the guidance and the experience of Daun Langston, Thomas Dineen, Tom Mathey, and many others. The members of the IEEE 802.3ah Task Force management and editorial team have helped me understand the process and make my contribution as effective as possible. I have shared many memorable moments with the representatives of other VDSL Alliance companies during our efforts to get DMT-VDSL selected as the technology for 10PASS-TS.

Many people have indirectly contributed to my decision to write a book about Ethernet in the First Mile. This includes my family, with their immense love of books, and the people around me who have first brought me in contact with computers and modems a long time ago. I'd like to thank Ed Eckert for his help in getting this book project going. Howard Frazier, John Cioffi, and Sam Lucero have kindly given me permission to quote them in this book. Special thanks to Steve Chapman of McGraw-Hill, for giving me the opportunity to publish this book and guiding me through the process, to Patricia Wallenburg, for getting the book into its current shape, and to the many other people involved in its production.

Finally, I would like to thank my wife Kathrin for supporting me in the time-consuming process of writing this book. I wouldn't have made it without her.

MICHAEL BECK
Antwerp

Acronyms

This is a book on data communications, all right! There's no way to avoid using tons of acronyms.

AIS	Alarm Indication Signal
ARP	Address Resolution Protocol
ATIS	Alliance for Telecommunications Industry Standards
ATM	Asynchronous Transfer Mode
AUI	Attachment Unit Interface
BER	Bit Error Ratio
CA	Connectivity Association
CC	Continuity Check
CFI	Call for Interest
CFM	Connectivity Fault Management
CO	Central Office
COL	Collision
CPE	Customer Premises Equipment
CRC	Cyclic Redundancy Check
CRS	Carrier Sense
CSMA/CD	Carrier Sense Multiple Access with Collision Detection
DES	Data Encryption Standard
DHCP	Dynamic Host Configuration Protocol
DMT	Discrete Multi-Tone
DSLAM	Digital Subscriber Line Access Multiplexer
DSM	Dynamic Spectrum Management
DTE	Data Terminal Equipment
EAP	Extensible Authentication Protocol
EFM	Ethernet in the First Mile
EPON	Ethernet Passive Optical Network
ESHDSL	Enhanced Single-Pair High-Speed Digital Subscriber Line

EVC	Ethernet Virtual Connection
FEC	Forward Error Correction
FTTH	Fiber to the Home
GBIC	Gigabit Interface Converter
GDMO	Guidelines for Definitions of Managed Objects
GMII	Gigabit Media Independent Interface
HDLC	High-Level Data-Link Control
IEEE	Institute of Electrical and Electronics Engineers
IETF	Internet Engineering Task Force
IFS	Inter-Frame Spacing
IP	Internet Protocol
ISS	Internal Service Sublayer
ITU	International Telecommunication Union
LAN	Local Area Network
LLC	Logical Link Control
LLDP	Link Layer Discovery Protocol
LMSC	LAN/MAN Standards Committee
MA	Maintenance Association
MAU	Medium Attachment Unit
MAC	Media Access Control
MCM	Multi-Carrier Modulation
MDIO	Management Data Input/Output
MDU	Multi-Dwelling Unit
MEF	Metro Ethernet Forum
MEP	Maintenance End Point
MIB	Management Information Base
MII	Media Independent Interface
MIP	Maintenance Intermediate Point
MMF	Multi-Mode Fiber
MPCP	Multi-Point Control Protocol
MPLS	Multi-Protocol Label Switching
MSTP	Multiple Spanning Tree Protocol
NAT	Network Address Translation
OAM	Operations, Administration, and Maintenance
OLT	Optical Line Terminal
ONU	Optical Network Unit
OSI	Open Systems Interconnection
PAM	Pulse Amplitude Modulation
PAR	Project Authorization Request
PCS	Physical Coding Sublayer
PDU	Protocol Datagram Unit

PHY	Physical Layer entity sublayer
PMA	Physical Medium Attachment (sublayer)
PMD	Physical Medium Dependent (sublayer)
PMS-TC	Physical Medium Specific–Transmission Convergence
PPP	Point-to-Point Protocol
PPPoE	Point-to-Point Protocol over Ethernet
PTM	Packet Transfer Mode
QAM	Quadrature Amplitude Modulation
RFI	Radio Frequency Interference
RFI	Remote Failure Indication
RSTP	Rapid Spanning Tree Protocol
SA	Security Association
SC	Secure Channel
SCB	Single-Copy Broadcast
SCM	Single-Carrier Modulation
SDH	Synchronous Digital Hierarchy
SDO	Standards Development Organization
SHDSL	Single-Pair High-Speed Digital Subscriber Line
SNAP	Subnetwork Access Protocol
SMF	Single Mode Fiber
SNR	Signal-to-Noise Ratio
STP	Spanning Tree Protocol
TC-PAM	Trellis Coded-Pulse Amplitude Modulation
TDM	Time Division Multiplexing
TDMA	Time Division Multiple Access
TLV	Type, Length, Value
TPS-TC	Transport Protocol Specific–Transmission Convergence
UNI	User-to-Network Interface
UPBO	Upstream Power Back-Off
VCSEL	Vertical Cavity Surface Emitting Laser
VDSL	Very-High-Speed Digital Subscriber Line
VLAN	Virtual (Bridged) Local Area Network
XID	eXchange IDentification

1

Introduction

*"Standards and sausages should not be
watched being made..."*   —ANONYMOUS

Introduction to Standards Development

NOTE TO THE READER: If you are only interested in the technical aspects
of Ethernet in the First Mile (EFM), feel free to skip this section. If you
care about the "why" and the "how" of this standard and others, however, you may find this section useful.

The Purpose of Standardization

The solution described in a technical standard is seldom the only solution to the problem it addresses. In fact, it often is not even the best
solution to the problem (complex engineering problems tend not to have
a "best" solution, but only a least inconvenient trade-off). All standards
do is identify a single point in the solution space of the problem, with
the goal of establishing a benchmark for interchangability and interoperability. This is all but true in the case of the EFM standard.

Standards differ from implementers' guides and books like this one
in that they only try to specify as little as possible to reach this goal,
thus leaving room for innovation and differentiation in implementation. As a result, standards for communication systems—including
EFM—will often specify the transmitter in great detail, while being
much less strict about the receiver; specifying both to the same extent
would be redundant, could introduce unintended contradictions, and
would unnecessarily limit the implementer's options. A lesser degree

of detail in the specification gives the implementer the freedom to build a receiver that operates correctly even when dealing with a non-compliant transmitter, following the principle of being "conservative in what you do, [and] liberal in what you accept from others" (a principle of robustness introduced in RFC 793).[1]

Interchangability allows users of a standard-compliant product to replace this product with another standard-compliant product (possibly from a different vendor) and be certain to get the exact same behavior. In this way, standards reduce the risk associated with investing in equipment; even if the first vendor goes out of business or discontinues the product, units from a different source can safely be used as a substitute. Standard-compliant products thus obtain a competitive edge over similar products that don't comply to any standard, unless a specific noncompliant product has established such a large market share that it has become a *de facto standard*.

Interoperability allows users of a first type of standard-compliant product to "operate" these in conjunction with a second type of standard-compliant product, even though the two types of products may be supplied by different vendors (think of a nut and a bolt, a VCR and a video cassette, a power plug and a wall outlet, etc.). In the case of a data communication system, one user's equipment may have to communicate with another user's equipment, of which the first user doesn't know any specifics other than that it complies to a certain standard. This is a second way in which standards reduce the risk associated with an investment, although a certain degree of confidence is required that the equipment with which interoperability is desired will be likely to be compliant to the appropriate standard.

As an example from the computer networking world, your high-end laptop may come equipped with a built-in 10/100/1000BASE-T port (i.e., a network interface card, which complies to these specific port types defined in IEEE Std. 802.3), and this is a low-risk choice for today's home and office network environments; the equipment that you will want to connect your laptop to—hubs, switches, and routers—is very likely to have ports that interoperate with the one on your laptop. However, if you have a goal of connecting this laptop to an ADSL modem that only has a 25.6 Mbps ATM Forum interface, you're out of

---

[1] Throughout this book, standards, recommendations, and RFCs will be referenced by the number assigned to them by the developing organization (optionally including the year of publication, if a specific edition is being referenced). You can find a list of all referenced standards, and information on how to obtain them, in the bibliography at the end of this book.

luck: Wrong standard—the network interfaces on the two machines will not interoperate (but the cable will fit, because the same connector standard is used!).

In order for the interoperability properties of a standard to actually give a product a competitive edge, the standard must be sufficiently successful in the market segment in which it operates. Your betamax VCR may be a wonderful machine and it will allow you to exchange cassettes with all your friends that have betamax VCRs, but that won't be much use to you if all your friends, who VHS machines! It is hard to predict which standards will be successful, but it is safe to assume that vendor buy-in, technological superiority, broad availability, and low cost of the product are all factors in the success of a standard. Sadly, if you study past standards for a little while, you will find that any rule of thumb to predict the success of a standard is bound to have a long list of exceptions. To make things worse, even in the face of *de jure*[2] *standards*, many markets exist in which the dominance of a single vendor will essentially set a *de facto standard* for a product, to which other vendors must comply in order to obtain or maintain market share.

IEEE 802.3 standards have a tendency to focus the market on a single solution. This is why 802.3 often takes on projects in areas where other (*de facto* or *de jure*) standards already exist, and sometimes ends up referencing an existing standard with minimal changes. The IEEE 802.3 stamp of approval and the "Ethernet" brand name go a long way in making the solution successful.

For the reasons of competitiveness outlined above, and sometimes to comply to external regulations, vendors will often choose to make products that comply with a number of applicable standards. As product development is often concurrent with standards development, many corporations understand the necessity of participating in the standards development process; it is the only way of keeping future products aligned with the standard, either by influencing the development of the standard directly or by intervening in the development of the product based on knowledge of the evolution of the draft standard. In this way, the first products that are (nearly) standard compliant usually hit the market by the time the standard approaches completion. The market advantage gained from such an early entry easily

---

[2] These are "official" or "legally enforced" standards, developed by a government-accredited standards development organization (SDO).

outweighs the manpower and travel cost associated with participation in standards development.[3]

Another way in which standards lead to profits for the companies participating in their development is through intellectual property rights (IPR) licensing. If a company owns a patent on a method or device of which the use is made mandatory or unavoidable by a standard, it could in theory sue any other company claiming compliance with the standard, for infringement of the patent. In practice, SDOs require that participants declare any relevant IPR prior to the final approval of the standard, and provide a license either for free or under reasonable and nondiscriminatory terms and conditions to anyone who wishes to use the patent to comply with the standard under development. Still, even under these conditions, a patent on an essential part of a successful standard can be a significant generator of revenue.

If an essential patent for which there is no declaration of nondiscrimination is discovered after the approval of a standard, this may seriously jeopardize the success of that standard, and may even lead to the withdrawal of the standard by the responsible SDO. The importance of the patent declaration policy is further illustrated by several recent court cases (described in [31]), in which the validity of standard-related patent infringement claims was disputed on the ground that the patent was not duly declared during the development of the standard to which it applied. In *FTC v. Dell Computer Corporation*, Dell was prohibited from enforcing its patent rights vis-à-vis implementations of a Video Electronics Standards Association (VESA) standard, for failing to declare its patents during participating in the standard's development. In *Townshend v. Rockwell International Corporation*, Townshend's patent rights in 56 kbps voice-band modems standardized in ITU were upheld, because Townshend had indicated its willingness to negotiate licensing terms on a reasonable and nondiscriminatory basis. At the time of writing, the case of *Rambus v. Infineon*—in which Rambus sued Infineon for infringement of its DDR SDRAM patents—is still in progress and is being followed with great interest by the industry. In 2003, the Federal Court of Appeals overturned an initial fraud verdict against Rambus which was based on Rambus' alleged non-compliance with IPR declaration obligations.[4] In

---

[3] An impressive plea for involvement in standards development can be found in section 16.10 of [39].

[4] *Rambus Inc. v. Infineon Technologies AG, Infineon Technologies North America Corp. and Infineon Tehchnologies Holding North America Inc.*, United States Court of Appeals for the Federal Circuit 01-1449, -1583, -1604, -1641, 02-1174, -1192.

this case, the vagueness of JEDEC's (the concerned SDO) IPR declaration policy at the time, and the fact that Rambus only filed the disputed patents after ceasing its participation in JEDEC, worked in
favor of Rambus.

The Lifecycle of an IEEE Standard

The development of IEEE standards is governed by the bylaws and
operating rules of the IEEE Standards Association (IEEE-SA). The
drafts are developed under the supervision of a sponsor organization,
which typically adds another set of rules to the game.

The LAN/MAN Standards Committee (LMSC) of the IEEE
Computer Society, a.k.a. IEEE 802, is the sponsor for the development
of standards pertaining to Local and Metropolitan Area Networks. It
is organized in a number of Working Groups (currently 22), which
develop standards in well-defined areas. Working Group 802.1 deals
with higher-layer interfaces and network architecture issues such as
bridging, security, and fault management. Working Group 802.2 is
responsible for the Logical Link Control protocols (see below). Media
Access Controls (MACs) and physical layer entity sublayers (PHYs) for
specific media and application sets are specified by the other Working
Groups, of which Working Group 802.3 deals with networks using the
Carrier Sense Multiple Access with Collision Detection (CSMA/CD)
method.

The structure of the LMSC was originally set up to allow Ethernet
(CSMA/CD) and Token Ring standards to co-exist. Today, Ethernet has
distinctly won the battle for the wired LAN market from Token Ring,
but the structure of the LMSC has allowed many other network technologies to be standardized efficiently and become successful in the
market by creating a symbiotic ecosystem of standards. Examples are
the IEEE 802.11 family (wireless LANs, also known as WiFi), and the
IEEE 802.16 (wireless access networks, also known as WiMAX).

The life of an IEEE standard development project starts with the
approval by NesCom (the New Standards Committee of the IEEE-SA)
of a Project Authorization Request (PAR). In IEEE 802, PARs are typically prepared by a Study Group formed as a result of a successful
Call for Interest (CFI) within a Working Group. Once the PAR is
approved, the Study Group is transformed into a Task Force with a
mandate to develop a draft standard. All the work of the Task Force is
subject to approval by the Working Group under which it resides.

As a result of the hierarchical organization of the LMSC in Working Groups and Task Forces, the actual approval of an IEEE 802 standard has the following three major phases:

Task Force Review. The initial work of the Task Force consists of transforming an adopted "baseline proposal" into text (this is the responsibility of the Editor). Once the initial text is available, a cycle of comment submission, comment resolution, and updating of the draft is iterated until the entire draft carries the consensus (see below) of the Task Force.

Working Group Ballot. The Working Group Ballot is conducted among all members of a Working Group. The subject of the ballot is the decision to forward the draft to the LMSC for ratification by means of a Sponsor Ballot. The process to get consensus for this decision is identical to the Task Force Review and Sponsor Ballot processes: A cycle of comment submission, comment resolution, and updating of the draft is iterated until the entire draft carries the consensus (see below) of the Working Group.

Sponsor Ballot. At the conclusion of the Working Group Ballot, a Sponsor Ballot Group is formed of members of the IEEE-SA that signed up for the Ballot Pool on this topic. A cycle of comment submission, comment resolution, and updating of the draft is iterated until the entire draft carries the consensus (see below) of the Sponsor Ballot Group.

At each of the steps outlined above, the project may die if a 75 percent approval rate is not met. At the conclusion of Sponsor Ballot, the IEEE Standards Board checks if the process was conducted correctly, and publishes the standard.

I have often been asked at what stage of this process the draft can be considered "stable." This question is of course of importance to implementers that want to have their product ready by the time the standard is published. Unfortunately, the only safe answer to this question is: "The draft is stable when the standard is published."[5] Every single step in the process can introduce changes to the draft. Note, however, that the Working Group ballot is only conducted when the draft is "technically complete," and that the key elements approved

---

[5] And even then, amendments, revisions, and errata may change the text of an approved standard!

by the Task Force are usually left unchanged (but see the section "Tones, tones, tones..." in Chapter 4 for a cautionary tale).

The standards created according to the process outlined here may be independent documents, revisions of existing standards, or amendments to existing standards.[6] The CSMA/CD Working Group maintains one standard,[7] IEEE Std. 802.3, to which all current projects (including EFM) are amendments. The process of "maintenance" deals with correcting any ambiguities, errors, or mistakes in approved standards. Maintenance requests are also balloted by the Working Group.

Whenever three or more amendments have been approved against a standard that is three years old, a revision of the complete document must be initiated before further amendments can be approved. A revision of IEEE Std. 802.3 was started upon completion of the EFM project. This revision consists of the consolidation of the latest amendments into a new edition of the base document, and resolving any comments submitted against any part of the standard in the course of the balloting process.

Getting Your Copy of the EFM Standard

The EFM standard was developed as an amendment to the base standard (IEEE Std 802.3-2002, including previous amendments for 10 Gigabit Ethernet, Data Terminal Equipment Power via Media Development Interface, and maintenance). This amendment is a stand-alone document containing the full text of all clauses created by the EFM Task Force (Clauses 56 through 67 and corresponding annexes, and Annex 4A) and change instructions for existing clauses (Clauses 1, 22, 30, and 45).

The amendment was published as IEEE Std. 802.3ah-2004. Several draft versions of this document have been made available outside the EFM Task Force (it is an IEEE 802 custom to offer drafts submitted for Working Group ballot for sale to the public, to encourage implementers to start early). These drafts are clearly marked as being temporary documents, subject to change, and must no longer be used except perhaps for purposes of historical research.

---

[6] As a rule in the LSMC, independent documents are designated by uppercase letters following the Working Group number, while amendments are designated by lowercase letters.

[7] Strictly speaking, there's also IEEE Std. 1802.3-2001, "Conformance Test 10BASE-T," which has been stable since 2001, as well as the Policies and Procedures and the Network Systems Tutorial.

As a result of the revision process that was started in 2004, a new edition of the base standard is being published in 2005, consolidating the amendments approved since the publication of the 2002 edition. It is strongly recommended to use the new edition of the base document (IEEE Std. 802.3-2005) rather than the stand-alone EFM amendment, as it contains a few corrections to the original text.

All published standards that are in force (whether they are amendments or integrated standards) can be purchased from IEEE. Documents older than 6 months can be obtained free of charge through the "Get IEEE 802" program.

Promoting Standards

Ethernet is a *disruptive* technology, in the sense that it tends to find application spaces where well-established high-quality solutions already exist and displaces these solutions with its own paradigm of simplicity and low cost (this observation builds on the concept of disruptive technological change as developed in [14]). At first sight, this displacement comes with a certain decrease in quality, but the disruptive technology offers its own way to circumvent the resulting problems: In the case of Ethernet, any quality-of-service problem can be solved by throwing more (cheap) bandwidth at it. Ethernet has successfully taken over the LAN space from FDDI/Token Ring (10 Mbps through 1 Gbps), and it is currently threatening SDH/SONET in the telecom core network (10 Gigabit Ethernet) and ATM in Internet access and aggregation networks (EFM and related standards). While the simple solution gains ground, the technology is enhanced to cover the features of the displaced technology. The cycle may continue to a point where the disruptive technology becomes mainstream, and then becomes the target of newer, more disruptive technologies.

Although any cheaper technology is inherently more attractive than a more expensive one, disruption can only occur when the customers can be convinced that the new technology will work just as well as the one they are used to. The standardization process by itself does not offer any guarantees to this effect. For this reason, vendors sometimes create alliances to promote new technology and accelerate the adoption of new standards. The EFM Alliance[8] was created in 2001 to promote the EFM standard through white papers, demonstrations, and interoperability events. It is not an SDO, nor does it participate as an entity in standards development.

---

[8] See www.efmalliance.org.

A Framework for the Definition of Data Networks

OSI and IEEE 802

The well-known Open Systems Interconnection (OSI) reference model provides a standardized framework for the definition of data networks (described in ISO/IEC 7498-1). The model's main strength is that it divides every communication system on the network in seven *layers*, each depending on the underlying layer and providing a service to the layer above (see Figure 1.1).

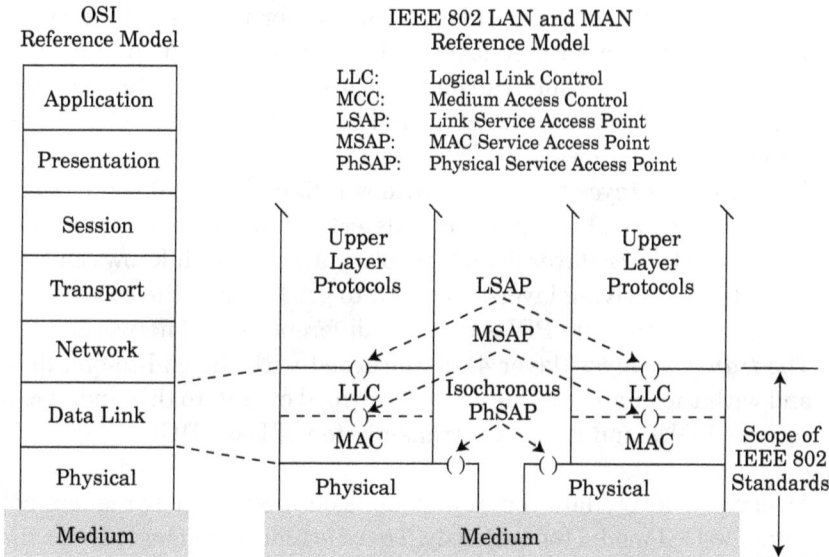

Figure 1.1 The ISO Open Systems Interconnection reference model.[9]

Each layer in the system talks to its peer in the other system(s), by passing protocol data units (PDUs) down the layer stack. PDUs received from lower layers are sent up the stack until they reach the intended layer. The model implies that the information available to each layer is limited to what is communicated through the interfaces with the immediately neighboring layers.

Moving down the stack, PDUs receive additional encapsulation (headers and trailers), which tells the receiving peer layer what to do with the PDU and where to send it; this encapsulation is removed by

---

[9] From IEEE Std. 802-2001, "Overview and Architecture," by IEEE. © 2001 IEEE. All rights reserved.

the peer (decapsulation) as the PDU moves up the stack at the receiving end.

The layers are defined in terms of the functions they perform. Different layers may be specified and implemented independently, as long as they offer the expected services through a commonly agreed interface. A detailed description of the functions of each layer can be found in any post-1984 book on the principles of data communication.[10] For the purposes of this book, it suffices to remind the reader of the main responsibilities of the lower four layers.

- The *physical layer* (layer 1) is responsible for moving bits across a *medium* (sometimes called layer 0), by transforming them into physical symbols suited for transmission over that particular medium (e.g., voltage/current waveforms on a copper wire, light pulses in an optical fiber, radio waves in free space).
- The *data link layer* (layer 2) provides reliable[11] logical connections between stations, by adding error detection and control overhead.
- The *network layer* (layer 3) adds topology awareness; it knows enough about the underlying layer-2 network to get PDUs to the correct destination, and to route PDUs between different layer-2 networks.
- The *transport layer* (layer 4) is concerned with the end-to-end link, and with the quality of the service provided over it; to this end, it can reorder PDUs and request retransmission of lost PDUs.

Modern books on data communications often warn their readers not to take the OSI model too seriously. There is indeed no magic about the number of layers defined in the OSI model, nor about layers' names or the specific way in which tasks are assigned to one layer or another. The model is inadequate for the description of label-switched networks (in an IP-over-MPLS-over-Ethernet protocol stack, MPLS is considered to be a "layer-2.5 protocol"), and it is too rigid to accommodate multiple occurrences of the same protocol in the protocol stack (e.g., Ethernet-over-MPLS-over-Ethernet).

This book takes a pragmatic approach, borrowing from the OSI model what is useful for our own purposes: the powerful concept of layering and sublayering—precisely defining the interfaces between functions, so every function of a "sublayer" knows exactly what to expect

---

[10] See for example [37] and [25].
[11] This does not mean that layer 2 has to assure delivery of all datagrams, but that delivered datagrams are correct.

from its neighbors—and an upper bound on the scope of the protocols defined in the IEEE 802 family of standards.

The standards developed by the LMSC concern the physical and data link layers. The overall network architecture governing the LMSC standards is contained in IEEE Std. 802. In the case of IEEE Std. 802.3 (Ethernet), the PHY, the optional Media Independent Interface and the Reconciliation Sublayer correspond to the OSI physical layer. The MAC sublayer, the optional MAC Control sublayer, and the optional Operations, Administration, and Maintenance (OAM) sublayer (see Chapter 8) correspond to the OSI data link layer. The Logical Link Control (LLC) sublayer defined in IEEE Std. 802.2 may optionally be added as the uppermost sublayer of the OSI data link layer.

Presently, the role of layer 3 is usually taken up by the Internet Protocol (IP), a connectionless protocol using a 32-bit address space (IPv4) or a 128-bit address space (IPv6), along with a set of associated control and routing protocols. The Transport Control Protocol (TCP) and the User Datagram Protocol (UDP) are two alternative protocols for the layer-4 transport functions, with and without frame retransmission, reordering, and rate control, respectively. All these layer 3 and 4 protocols are maintained by the Internet Engineering Task Force (IETF). In the world of TCP/IP, the top three layers are commonly considered together as one extended application layer.

IEEE Std. 802.3 adheres strictly to the layering paradigm. Within the physical and data link layers, *ad-hoc sublayers* are defined for the purpose of reuse and interchangeability between different transceiver types and to create a logical partitioning of management objects. These sublayers may introduce additional encapsulation and decapsulation.

The Three Faces of the Ethernet Frame

As a result of its origin as a proprietary technology, Ethernet supports three different variants of the Ethernet frame. It is important to note that all three are compliant with the Ethernet standard, and that they can coexist on the same network infrastructure.

The type/length field is often a source of confusion. The original DIX specification used this field as a type field. In the early 802.3 work, where the presence of the 802.2 LLC sublayer was presumed, there was no more need for a type field at the MAC level, and the field was redefined as a length field. This distinction is still often quoted erroneously as *the difference between Ethernet v2 and IEEE 802.3*. With the introduction of MAC Clients other than LLC, the IEEE 802.3 length field got its original type interpretation back for values in excess of 1536. The alleged distinction between Ethernet v2 and IEEE 802.3 thus vanishes, as both interpretations are legal in the current version of IEEE Std. 802.3.

Ethernet v2 encapsulation. The most straightforward encapsulation method is the one originally used by DEC, Intel, and Xerox (DIX) in their Ethernet Version 2 specification. It uses a 2-byte "type" field to indicate the nature of the MAC Client protocol (the "Ethertype"). Currently, only Ethertype values above 1536 (decimals) are legal, to avoid interpretation as a "length" field (see below). Type field values, or "Ethertypes," are assigned by the IEEE Registration Authority and published by the Internet Assigned Numbers Authority (IANA).[12]

Encapsulation with a type field is used for Ethernet-specific protocols such as MAC Control (Ethertype 0x8808) and Slow Protocols (0x8809—this includes the Link Aggregation Control Protocol and OAM). It is also commonly used to carry layer-3 protocols such as IPv4 (0x0800) and IPv6 (0x86DD).

Specific Ethertypes are used to indicate that a frame is Virtual Bridged Local Area Network (VLAN) tagged (see Chapter 10) or belongs to an end-to-end OAM flow (this use is still under discussion; see also Chapter 8).

Ethernet with LLC encapsulation. IEEE Std. 802.2 specifies the LLC sublayer, an optional sublayer on top of the MAC sublayer, together forming the Data Link layer of the OSI reference model. The LLC sublayer provides different service types:

Type 1 operation. Connectionless operation; PDUs are not acknowledged, and there is no flow control or error recovery.

Type 2 operation. Connection-oriented operation; PDUs are acknowledged by the recipient.

Type 3 operation. Connection-less operation; command PDUs are acknowledged by specific response PDUs.

Type 2 operation is rarely used on Ethernet networks, as the most common clients running over Ethernet do not require the connection-oriented service. We can also safely skip the discussion of Type 3 oper-

---

[12] For a list, see www.iana.org/assignments/ethernet-numbers.

ation, as this type relies on the MA-UNITDATA-STATUS primitive, which is not implemented by the Ethernet MAC. In this book, we will always assume Ethernet to be connectionless and unacknowledged (Type 1 operation).

The LLC encapsulation, shown in Figure 1.2, provides an indication of the source and destination service access point, removing the need for a separate type field in the MAC frame. Hence, the type field of the Ethernet frame is used as a length indicator (values up to 1522 bytes) in case the LLC sublayer is present.

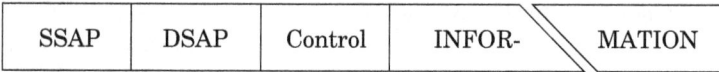

SSAP	DSAP	Control	INFOR-	MATION

Figure 1.2   LLC PDU format.

The presence of the length field removes the minimum payload size requirement that Ethernet places on its client. Without LLC, the payload of an Ethernet frame needs to be at least 46 bytes to get to the minimum Ethernet frame size of 64 bytes after adding the Ethernet overhead. With LLC, and with a length field in the Ethernet header, shorter payloads can safely be padded to the required length, without ambiguity about the boundary of the original payload.

Half of the link service access point (LSAP) address space is administered by the IEEE Registration Authority,[13] while the other half can be used freely by network administrators. Ethernet with LLC encapsulation is the preferred protocol stack according to the IEEE 802 architecture, as it brings a uniform data link service definition to the diverse physical layer technologies defined in the IEEE 802 family of standards. It is mandatory for bridge-specific protocols such as the Spanning Tree Protocol and the Generic Attribute Registration Protocol (both use SSAP = DSAP = 0x42). LLC encapsulation can also be used with any layer-3 protocol, IP (SSAP = DSAP = 0x06) being the most common choice, but this is not required for compliance with IEEE Std. 802.3.

Ethernet with LLC and SNAP encapsulation. The 1-byte DSAP field available in the LLC header to indicate which layer-3 protocol is being used by the Link Service client causes two serious limitations:

---

[13] For a list, see www.iana.org/assignments/ieee-802-numbers.

- The address space for protocols that do not have a reserved LLC address is relatively small.
- Interworking with LAN technologies that use a 2-byte Ethertype is impossible.

Subnetwork Access Protocol (SNAP) provides a way to remove these limitations: It adds another 5-byte header, which consists of a 3-byte Organization Unique Identifier (OUI) and a 2-byte protocol identifier. Existing Ethertypes can be used as protocol identifiers when the OUI is set to all-zeros. The presence of the SNAP header is indicated in the LLC header by setting the LSAP and LDAP to "SNAP" (0xAA).

Incidentally, LLC and SNAP are used as an adaptation layer to carry packet-based protocols over Asynchronous Transfer Mode (ATM) networks. In that case, the payload of the LLC/SNAP container is often again an Ethernet frame (this is the way Ethernet is currently transported over ADSL links).

The Ethernet in the First Mile Project

A Call for Interest on "Ethernet in the Last Mile"

In November 2000, the CSMA/CD Working Group of the LMSC hosted a Call for Interest on the topic of Ethernet in the Last Mile (ELM).[14] The goal of the presenters was to start a project on Ethernet in Subscriber Access Networks. The scope was understood to include public access networks as well as private networks with access characteristics: university campuses, hotels, hospitals, business centers, etc. Finally, multi-dwelling units (MDUs) and multi-tenant units (commonly referred to as MxU) were agreed to be in scope, with the exception of the home network, which is completely managed by the end user (typically the realm of Ethernet networks over structured wiring, WiFi, and HomePNA).

The topologies on which the group eventually agreed to work were point-to-point copper, point-to-point fiber, and point-to-multipoint fiber (PON). Each of these topologies would be addressed by one or more "port types." A port type is a coordinated set of specifications for transceivers achieving a given performance on a given medium; the specification should be sufficiently detailed to guarantee interoperability between independent implementations. Port types receive

---

[14] Before the end of the meeting, the name was changed to Ethernet in the First Mile (EFM), for reasons explained later.

names such as 10BASE-T and 100BASE-TX, giving an indication of bitrate, modulation, medium, and (optionally) an additional differentiator.[15]

To start an IEEE 802 standardization project, the LMSC requires five criteria to be addressed: distinct identity, broad market potential, economic feasibility, technical feasibility, and compatibility with the existing IEEE 802 architecture.

The easiest way to demonstrate technical feasibility is to point at an existing system that does exactly that which the proposed standard aims to achieve. This sometimes happens, and when it does, it should trigger an important question: If a solution already exists for our problem, what is the use of specifying a postfactum standard for it? The answer must lie in the other criteria: The goal of the new standard must be to do the same thing for a broader market, in a cheaper way, and in accordance with the IEEE 802 architecture.[16]

The disadvantage of starting a standardization project on a subject that has already been proven technically feasible in this way is having to deal with companies (and other standards organizations!) that wish to see their solution standardized with as few changes as possible. Although this makes economic sense, this can lead to a total deadlock if several competing solutions exist, backed up by groups of equal strength. Even if such a situation is eventually resolved and a single solution is standardized, it may take a long time before prestandard solutions are displaced from the market.

Enter the Bell-Heads

In the particular case of EFM, the group's main division lay along the line that separates the so-called Bell-heads from the so-called net-heads. The former group, consisting of people with a background in public access networks (such as telephony and DSL), tends to think of Ethernet as another way of framing data. The latter group, including may people who spent years standardizing physical layers for local area networks, sees Ethernet as a way of life. (Of course, there are open-minded people in both these groups, without whom the EFM

---

[15] Note that 100BASE-X and 1000BASE-X are not names of port types. These terms designate families of port types sharing a common physical coding sublayer.

[16] The criterion of distinct identity only requires the project to be different from other IEEE 802 projects. It does not prevent projects being formed that do work in areas where other standards bodies are active.

standard would never have come this far! To be honest, I went in as a sort-of Bell-head, and came out as a sort-of net-head...).

Traditionally, most of the standardization work in the domain of access networks has been taking place in national or international standards bodies such as ITU-T Study Group 15, ATIS T1E1.4, and ETSI TM6; organizations whose work is based on a "no objections" model, and whose decisions are therefore characterized by compromise. By the time EFM came along, the copper access landscape had been neatly divided by these SDOs into a long-reach residential segment (ADSL), a long-reach business segment (SHDSL), a short-reach high bit rate segment (VDSL), with further subsegments under study (ADSL2, enhanced SHDSL). Even in the VDSL space, two different technologies, DMT and QAM, were allowed to survive, because no unanimity could ever be reached to drop either one.

The LAN people from IEEE 802.3 adhere to a 75 percent majority rule. This means that there is no need to strive for unanimous consent;[17] it is sufficient to reduce the group of opponents to 25 percent or less to get any technical motion passed. The next motion that gets approved might have a different minority opposing it, which implies that this system by itself does not guarantee consistency. Because a standard must be more than a set of individually approved ideas, a draft can only become a standard after a balloting process in which the document as a whole is evaluated and eventually approved.

What's in a Name?

From the point of view of a network service provider, the subscriber access network is the last part of the network that data have to traverse in order to reach the customer; hence, it is normally referred to as the last mile. At the end of its first meeting (a call for interest launched under the name "Ethernet in the Last Mile"), the Study Group intentionally turned this convention around and started using the term "first mile" for the access network to convey the impression that the network would now be looked at from the consumer's point of view. However, the access network has and always will have two ends, and the actual customers for most of the EFM equipment will be network service providers, as these providers often own the "customer premises equipment" (CPE) as well as the central office (CO) equipment.

---

[17] I am purposefully avoiding the word "consensus" here, as it is defined differently by different organizations.

A lot of discussion went into the name of the project. Some feared that going against a well-established convention would lead to confusion among customers. Others claimed that the "first mile" approach was bound to catch attention more easily, and that the equivalence between the two terms would be obvious to everyone. In my personal experience, I've never come across a customer who didn't get the pun, or an audience that didn't smile when I started a presentation with a slide reading "FIRST MILE = LAST MILE."[18]

For the IEEE 802.3 Working Group, the subscriber access network really was the last mile to be conquered. After all, Ethernet is already present in all other segments of the wired network.

The Need for Copper

With the exception of certain MDUs equipped with structured wiring that supports 10BASE-T or 100BASE-T for in-house distribution of data, typical subscribers do not have direct access to a medium that supports Ethernet between their home or office and a service provider. What they do typically have access to is a phone line, which is actually quite well suited for moderate-speed data transport, as proved by the current success of ADSL.

From the point of view of some xDSL standardization experts, EFM over point-to-point copper was the single most useless standardization effort ever undertaken. The only thing needed to allow for the transport of Ethernet frames (see Table 1.1) over xDSL links is an adaptation layer between Ethernet's Media Independent Interface (MII) and the packet-based γ-interface common to the ITU-T's xDSL specifications. Instead, the EFM Task Force chose to specify two complete physical layers for operation over point-to-point copper, both based on existing xDSLs.

Useless or not, the EFM project assisted in breaking the deadlock that VDSL standards were in with respect to the line code; and even after the recent progress in ITU-T, 2BASE-TL, and 10PASS-TS, remain the *only* standard xDSL flavors that are internationally usable.

---

[18] The credit for this idea goes to Michael J. Bennett, who suggested it in a message on the EFM email reflector on April 2, 2001.

TABLE 1.1   The IEEE 802.3 MAC Frame or "Ethernet" Frame

Field	Length	Description
Preamble	7 bytes	Fixed pattern of 0x55 bytes to facilitate synchronization at the receiver
Start Frame Delimiter (SFD)	1 byte	Fixed field 0xE5 to indicate the start of the data frame
Destination Address	6 bytes	Unique MAC address of the receiving station
Source Address	6 bytes	Unique MAC address of the transmitting station
Length/Type	2 bytes	Field indicating the content type in the payload field (value ≥1536) or the length of the frame (value ≤1500)
QTag Control Information (optional)	2 bytes	Used for VLAN-tagging. If present, the Length/Type field must be set to 0x8100 (QTagType), and the MAC Client Length/Type field must follow.
MAC Client Length/Type	2 bytes	Field indicating the content type in the payload field (value ≤1536) or the length of the frame (value ≤1500).
Payload	Up to 1500 bytes	Whatever data the MAC Client chooses to transmit, padded to a length of at least 42 bytes
Frame Check Sequence (FCS)	4 bytes	A 32-bit CRC to verify frame integrity

NOTE—According to the MAC frame description shown above, the *Length/Type* field is always present, and the *QTag Control Information* and *MAC Client Length/Type* fields are only present in frames of type *QTagType*. An alternative interpretation considers the *MAC Client Length/Type* field as the unchangeable companion of the payload, and the combination of *Length/Type* = 0x8100 and *QTag Control Information* as a "VLAN tag" inserted between the Destination Address and the *MAC Client Length/Type* field.

From the point of view of the IEEE 802.3 Working Group, the definition of new PHYs was a logical step. IEEE Std 802.3 reads like a catalog of port types and physical layer specifications for use on a variety of media and topologies, spanning a range of speeds from 1 Mbps up to 10 Gbps (currently). There was no PHY available for a single pair of voice-grade copper, so they made one. They referenced existing standards for obvious reasons and made some simplifications while they were at it.

From the point of view of a network equipment vendor with a proprietary solution to the single-pair voice-grade copper problem, a new standard close to the existing product would be a great competitive edge and a marketing tool. That explains why the EFM project received so much attention and support from the industry.

But why include new PHYs for single-pair voice-grade copper at all? Optical fiber is already the technology of choice for long-haul and high bit rate network segments. For shorter stretches and lower bit rates, such as typical subscriber links, fiber is the highest capacity and therefore the most future-proof medium. The optical fiber, by itself, is even rather cheap. However, the cost of transceivers, and especially the installation of the fiber, is often prohibitive for deployment except in greenfield deployments, where fiber can be installed during construction at a cost that is virtually the same as the cost for installing copper. This significant installation cost of fiber-to-the-home (FTTH) is a financial hurdle that may cause fiber deployments to take many years before reaching the ubiquity of copper networks.

There are of course two sides to a business model: cost and revenue. The attainable bandwidth of a fiber link is significantly higher than that of a copper pair. As a result, more services can be offered to the subscriber in a FTTH deployment, and more revenue can be generated. Video-on-demand, broadcast TV (regular or HDTV), video telephony, high-speed Internet access, on-line gaming, virtual LANs, and so on can all be offered via the same strand of fiber. If the subscriber is willing to pay enough for these services, the business case *will* work.

Given the economical difficulties inherent to deploying optical fiber all the way to the user, the typical transition scenario is that in which a cheaper technology is used in at least part of the first mile, a scenario known as fiber-to-the-cabinet/curb (FTTC). The existing copper pairs used for the plain old telephony service (POTS) are the obvious solution, and the success of ADSL shows that this solution is technically and economically feasible. A technology like VDSL or 10PASS-TS is needed to offer each subscriber a more substantial fraction of the bandwidth that is available at the cabinet. John Cioffi, a Stanford professor active in the field of xDSL, claims that copper access technologies will be around alongside fiber for the next 100 years or so.[19]

We must, however, be careful with our conclusions. With certain operators estimating that the deployment of FTTH may be *only* around 16 percent more expensive than FTTC [19] it seems to make economical sense to skip the FTTC stage in those markets in which FTTC is not expected to be around long enough to recover the cost associated with deploying a high-speed copper drop from the cabinet to the user.

---

[19] This statement was made at a workshop on emerging transmission technologies [15]. When I recently reminded Professor Cioffi of this statement, he reiterated that there are 96 years to go and the number of DSLs is still rising! (John Cioffi, private communication, January 2005.)

The figures submitted by the Task Force in March 2001 in support of the "Broad Market Potential" criterion (*"The available market is estimated by third-party analysts at greater than 40 million subscribers in the United States and 150 million subscribers worldwide by 2005."* [23]) seem overly optimistic today. A comparison of analyst's reports from November 2002 and August 2003, shows a significant downward correction in the forecasts: from 35 million subscribers worldwide in 2006 [29] to 24 million subscribers worldwide in 2007 [30] (both forecasts are significantly lower than the figures proposed by the EFM Task Force).[20] However, the author of the cited reports believes that the general trend toward Ethernet in the access network holds: DSLAMs are increasingly being backhauled by Gigabit Ethernet, VDSL (and its EFM counterpart 10PASS-TS) is increasingly being used in the first mile, and service providers worldwide are getting serious about FTTH (which eventually will mean Ethernet).[21]

As soon as it was decided that EFM was going to have a point-to-point copper PHY (at the second Study Group meeting—after an initial negative decision), it was clear that it was going to be very xDSL-like. The Task Force never set any objectives that went beyond what was already achievable with commercially available xDSL products, which is why they did not come up with anything significantly new.[22] This being said, it is probably a good thing for the industry that the Task Force had the wisdom to base its copper PHYs on pre-existing standards, rather than pre-existing nonstandard products. And in the process, the industry gained a well-defined and standardized way to transport Ethernet MAC frames over *any* standard xDSL.

The Limits of EFM

Project Scope

According to its full title, the EFM standard specifies "Media Access Control Parameters, Physical Layers, and Management Parameters

---

[20] For a correct interpretation of the figures, it must be noted that the cited reports define EFM as "the use of the Ethernet protocol in conjunction with various physical media—copper, coax, or fiber—to provide a broadband service link between a service provider's local exchange/central office/head-end and the subscriber's premise." This definition is broader than the port types specified in IEEE Std. 802.3ah-2004.

[21] Sam Lucero, private communication, December 2004.

[22] At some time in the process, there were certain operators who actually wanted to define something new, transporting higher bitrates over longer distances than currently possible ... at no extra cost! That was an engineering challenge that the Task Force was *not* willing to take on.

for Subscriber Access Networks." Except for forgetting to mention the new Operations, Administration, and Maintenance (OAM) protocol, this is an accurate description of what can be found in the standard: a set of PHYs optimized for the access network, optionally managed, and the tweaks to the MAC that are required to use them. The precise objectives that the Task Force adopted to get to this specification are shown in Annex A.

The access-specific problems were new to the IEEE 802.3 community. Some of the architecture-specific problems that the EFM Task Force faced, however, had already been solved in a more or less satisfactory way in other standards: HomePNA had already tackled the MAC-PHY rate matching issue, GPON already addressed the issue of accessing the shared fiber infrastructure, and DOCSIS had its own solution for segregating customer-domain spanning trees from provider-domain spanning trees.[23] Also at the PHY level, many standards existed that the EFM Task Force could "borrow" material from to solve its own problems. These standards are incorporated by reference in the EFM standard where appropriate.

The existing standards at the network level, on the other hand, were never considered to be in the scope of EFM. By construction, EFM is compatible with the IEEE 802 network view and the related standards of the IEEE 802.1 family, and no modifications were made to the (sub)layers above the MAC Service interface.

New PHYs for the Old MAC

The MAC sublayer specified in IEEE Std. 802.3 takes care of the contention issues that can arise when different clients attempt to access a shared transmission medium. In half-duplex mode, this is achieved by means of CSMA/CD, which includes a random back-off algorithm to resolve contention resulting from two stations attempting to access the medium at the same time. In full-duplex mode, there are exactly two stations on the local area network (LAN) that can transmit independently and simultaneously over two different (logical) paths.

The upper interface of the MAC sublayer, communicating with the MAC Client, is conveniently named the MAC Service. To ensure consistency across different MAC technologies—which is vital to the concept of MAC bridging, as explained in Chapter 10—the MAC Service

---

[23] This problem is *not* addressed by the EFM standard, but will be addressed by 802.1ad.

is defined in a separate document, ISO/IEC 15802-1.[24] The primitives of the MAC Service interface are shown in Table 1.2. Unfortunately, the nomenclature of the service primitives is not identical between ISO/IEC 15802-1 and the different IEEE 802 standards.

It is important to note that the MAC Service is not defined in terms of specific signals on an (electrical or other) interface, unlike most other named interfaces discussed in this book. The invocation of the MA-UNITDATA request primitive typically consists of a MAC Client (e.g., the IP stack in the operating system of an end station) copying a packet into the physical memory of the Ethernet device driver and calling the "start output" routine. The invocation of the MA-UNIT-DATA indication primitive consists of a "receive-complete" interrupt to the Ethernet device driver, triggering the latter to copy the packet from the Ethernet device into memory accessible to the MAC Client. (Chapters 1–4 of [47] provide an excellent overview of the interaction between an Ethernet device and an IP stack.)

TABLE 1.2   MAC Service Specification (802.1AC notation)

Primitive	Description	Parameters
MA-UNITDATA request	Primitive used by the MAC Client to transmit a MAC service-data-unit (MSDU) to a peer	Destination Address, Source Address, Routing Information, MSDU, Priority
MA-UNITDATA indication	Primitive used by the MAC to indicate to the MAC Client that a MSDU has been received	Destination Address, Source Address, Routing Information*, MSDU, Priority*

NOTES
1. The value of MA-UNITDATA indication parameters marked with an asterisk is not necessarily equal to the value provided in the corresponding MA-UNITDATA request.
2. If the Source Address parameter is not present in the request, the MAC sublayer inserts the unique device-specific MAC address into the frame (this is the normal mode of operation).
3. When Routing Information is included to perform source-routing (which is actually contrary to the Ethernet principles of operation), a Q-tag (see the section "VLAN Bridges" in Chapter 10) must be added to the frame with the CFI bit set.
4. The priority parameter is ignored by the Ethernet MAC. Priority-tagged Ethernet frames may, however, be queued according to priority in priority-aware MAC bridges. Quality-of-Service is thus observed at the MAC Relay level, not at the level of the Ethernet MAC itself.

---

[24] This standard is particularly hard to find. The reader may want to refer to IEEE Draft P802.1AC instead, a document that combines the specifications of ISO/IEC 15802-1 with the ISS definitions from IEEE Std. 802.1D.

There are two essential MAC Service primitives in Ethernet:[25] "MA_DATA.request" (equivalent to "MA-UNITDATA request") is asserted by the MAC Client when it wishes to transmit a datagram, and "MA_DATA.indication" is asserted by the MAC sublayer to indicate that a received datagram is available. The MAC Client can be the LLC sublayer or a layer-3 protocol, such as IP, which is most often implemented in software as part of the operating system environment of the host. Furthermore, tunnelling protocols—sometimes called "layer-2.5 protocols"—such as PPPoE and MPLS may be present between layer 2 (the Ethernet MAC and optionally LLC) and layer 3 (IP).

In MAC Bridges,[26] the so-called MAC Relay Entity connects to a different MAC Entity at each of its ports. This connection is provided by the Internal Service Sublayer (ISS) access point, of which the main difference with the MAC Service access point is the fact that it is *promiscuous*, i.e., it exposes *all* frames received from the medium, not just those destined to that individual MAC. In addition to the parameters present in the MAC Service primitives, a number of parameters with local significance are present at the ISS (frame_type, mac_action, and frame_check_sequence).

The lower interface of the MAC connects to the medium through a physical layer entity (PHY). With the introduction of "Fast Ethernet" (100 Mbps), a Reconciliation Sublayer (RS) and MII were introduced, which could be used to attach any 10 Mbps or 100 Mbps PHY to the MAC. In integrated solutions, the MII is usually not present as a physical interface, and when it is, it is often a nonstandard variant with reduced pin-count.

IEEE 802.3ah, the "EFM" Task Force, was based on a Project Authorization Request (PAR) in which the scope of the project is specified as "Define 802.3 Media Access Control (MAC) parameters and minimal augmentation of the MAC operation, physical layer specifications, and management parameters for the transfer of 802.3 format frames in subscriber access networks at operating speeds within the scope of the current IEEE Std. 802.3 and approved new projects." In this book, the term "EFM" will be interpreted according to the scope of the EFM Task Force. A lot of other issues need to be resolved if an Ethernet-based service is to be successfully deployed in a subscriber

---

[25] Two more primitives are available when the optional MAC Control sublayer is present: MA_CONTROL.request and MA_CONTROL.indication.
[26] See Chapter 10 for more details on bridges.

access network. This extended problem space is sometimes referred to as EFM as well, but it would be better to reserve a separate name for this, such as "Ethernet Subscriber Access."

The new EFM PHYs are added to the long list of PHYs already specified by IEEE Std. 802.3, and work with the same MAC (after applying the "minimal changes"). The EFM PHYs can be connected to existing standard-compliant implementations of the MAC, such as those implemented in MAC bridges (also known as "Ethernet switches") or network interface cards (NICs).

Economics

Ethernet technology resides in a virtuous circle of simplicity, popularity, and cheapness. Thanks to the enormous volumes of Ethernet components being produced and sold (95% of all LANs are believed to be Ethernet-based today), the price of equipment is low. Deployment of Ethernet networks is relatively easy, which explains the high demand for Ethernet equipment. Many data terminals (e.g., laptops, PCs) are shipped today with built-in 10/100/1000BASE-T Ethernet ports.

The prospect of using cheap and simple equipment in a subscriber access network is obviously very attractive, because it allows for higher margins. However, the simplicity of Ethernet in this context is deceiving. It is just as easy to build a simple Ethernet-based access network as it is to build a simple network based on any other protocol. None of these simple networks will be particularly suited to make you money. Ethernet was invented as a solution for LANs, which consist of users that belong to a single organization (e.g., a corporation, a home), and that don't intend to harm each other. Access networks consist of two classes of users: subscribers and service providers. The subscribers are not necessarily friendly towards the service providers or towards other subscribers. That simple fact creates a necessity for security and user traffic segregation. This means that much of the simplicity that made Ethernet successful is gone when you move to an access network. These issues were outside the scope of the EFM Task Force.

For the actual subscriber ports, on any given media, there will be no difference in cost between a port supporting native Ethernet frames and a port supporting Ethernet-over-ATM. The port types defined by the EFM project are not sufficiently different from existing ATM-based port types to cause an appreciable difference in cost. The only thing that could favorably influence the economics of EFM is the possible

mass adoption of broadband subscriber access due to the magical
power of the brand name "Ethernet."

The EFM Task Force eventually standardized eight PHYs (some of
them comprising a different port type for each side of the link), summarized in Figure 1.3. In addition to these EFM PHYs, 2PASS-TL is
also shown in Table 1.3, although it is not part of the standard. 2PASS-
TL is, however, essential for EFM-like deployment to residential subscribers, and will be covered in the last section of this book.

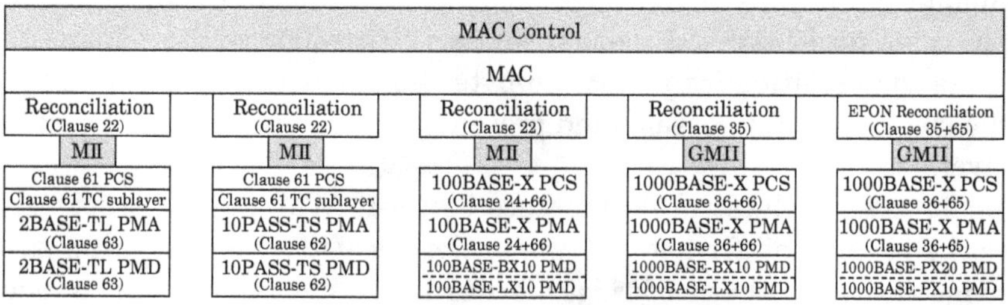

MAC Control				
MAC				
Reconciliation (Clause 22)	Reconciliation (Clause 22)	Reconciliation (Clause 22)	Reconciliation (Clause 35)	EPON Reconciliation (Clause 35+65)
MII	MII	MII	GMII	GMII
Clause 61 PCS	Clause 61 PCS	100BASE-X PCS (Clause 24+66)	1000BASE-X PCS (Clause 36+66)	1000BASE-X PCS (Clause 36+65)
Clause 61 TC sublayer	Clause 61 TC sublayer			
2BASE-TL PMA (Clause 63)	10PASS-TS PMA (Clause 62)	100BASE-X PMA (Clause 24+66)	1000BASE-X PMA (Clause 36+66)	1000BASE-X PMA (Clause 36+65)
2BASE-TL PMD (Clause 63)	10PASS-TS PMD (Clause 62)	100BASE-BX10 PMD / 100BASE-LX10 PMD	1000BASE-BX10 PMD / 1000BASE-LX10 PMD	1000BASE-PX20 PMD / 1000BASE-PX10 PMD

Figure 1.3   Overview of PHYs defined in IEEE Std. 802.3ah-2004.

TABLE 1.3   EFM PHY Overview

Name	Media	Bit Rate	Reach	Standard	Book
2BASE-TL	Voice-grade copper	2 Mbps	2700 m	61 + 63	Chapter 5
2PASS-TL	Voice-grade copper	2 Mbps	2700 m	n/a	Chapter 5
10PASS-TS	Voice-grade copper	10 Mbps	750 m	61 + 62	Chapter 4
100BASE-LX10	Dual SM fiber	100 Mbps	10 km	24 + 60	Chapter 6
100BASE-BX10	Single SM fiber	100 Mbps	10 km	24 + 60	Chapter 6
1000BASE-LX10	Dual SM fiber	1000 Mbps	10 km	36 + 59	Chapter 6
1000BASE-BX10	Single SM fiber	1000 Mbps	10 km	36 + 59	Chapter 6
1000BASE-PX10	SM fiber	1000 Mbps	10 km	36 + 58	Chapter 7
1000BASE-PX20	SM fiber	1000 Mbps	10 km	36 + 58	Chapter 7

Connections

Traditional xDSL-based and PON-based access networks in Europe and North America use ATM rather than Ethernet as their data link layer. ATM is a layer-2 and layer-3 technology, which transports data in fixed-sized cells rather than variable-sized packets. Each cell carries a 5-byte header followed by 48 bytes of payload. Obviously, an adaptation layer is required to encapsulate higher-layer protocol datagram units that are not exactly 48 bytes long; several such "ATM Adaptation Layers" (AAL) have indeed been defined. AAL5, for example, adds an LLC/SNAP header and a CRC trailer to the datagram, splits the resulting frame into 48-byte chunks, adds padding bytes to the last chunk if necessary, and presents the 48-byte units to the ATM layer.

The main architectural difference between ATM and Ethernet networks is that the former are *connection-oriented* while the latter are *connectionless*. Ethernet bridges forward each individual frame to one or more ports, based on the frame's destination address and the bridge's most current knowledge of the network topology. In an ATM network, an end-to-end *connection* is established before any data are transmitted. In the connection set-up phase, each ATM switch is told with which incoming and outgoing port each *virtual path* corresponds. This information is maintained for the duration of the connection. Beside the virtual path identifier, ATM cells also carry a virtual channel identifier in their header, which opens up the possibility of using the same virtual path, between the same end points, for multiple services.

The main advantages of the cell approach are the possibility to reserve resources during the connection set-up phase in order to guarantee a certain quality of service level during the connection, the possibility to multiplex different services onto the same physical link with a negligible latency penalty for the highest-priority service, and the possibility to inverse multiplex a stream of ATM cells over multiple parallel physical links (IMA, the ATM version of bonding). The disadvantage is simply the fact that connections need to be set up, which requires additional intelligence in the network elements along the path, and the resulting cost penalty.

For a first-mile technology, the advantages of ATM are needed most on low-bitrate links, such as *long* xDSL lines. It must be noted that PME aggregation as defined by EFM Copper presents a decent alternative to IMA, and that the recent introduction of pre-emption to the ITU-T G.992.5 variant of the EFM-TC sublayer takes care of the service multiplexing issue (hence the nickname of "ATM light" for this sublayer in ITU-T). In other words, there is no need to introduce the connection-oriented paradigm to the xDSL-link to make it useful. The EFM-TC sublayer is described in Chapter 3.

Multiple Protocol Label Switching (MPLS) offers a packet-based alternative for connection-oriented networking. As a layer-2.5 protocol, its header is placed between the Ethernet header and the IP header. The header carries a stack of *labels*, which indicate to subsequent MPLS switches to which port the packet shall be forwarded. With respect to ATM, the hassle of fragmenting the packet is removed, but the complexity of setting up a connection remains. The Point-to-Point Protocol over Ethernet (PPPoE) is another layer-2.5 protocol used to distinguish different IP flows sharing an Ethernet infrastructure, but it is not a truly connection-oriented protocol, as there are no "PPPoE switches" that use information in the PPPoE header for their forwarding decisions.

As the term "layer 2.5" implies, MPLS and PPPoE operate above layer 2, and hence outside the realm of Ethernet. Hence, the EFM project has never had the intent or the need to define any sort of connections in this sense. However, VLANs, discussed in chapter 10, can provide some of the benefits of connection-oriented networking, without going as far as providing switched virtual paths.

Deployment Scenarios

As explained above, EFM is no more than a set of PHY specifications (and some changes to the Reconciliation Sublayer, MAC, and MAC Control), which may be used to build an EFM modem. To get to a marketable service based on EFM, higher-layer protocols and network nodes are a necessity. Other SDOs and fora, such as the DSL Forum and the Metro Ethernet Forum (MEF), are focused on the definition of such services and the interfaces that characterize them. Research consortia such as the European MUSE consortium,[27] bring key players together to coordinate input to these fora. A description of these services and the way they are implemented can be found in the section "Ethernet Subscriber Access" in Chapter 10.

---

[27] www.ist-muse.org.

Chapter

2

EFM's New, Old MAC

"Ethernet, implemented by almost everybody in communications, adapts how data [are] sent as computing technology improves. It's simple, easy to use, and pervasive. You plug it in and it works." —Bob Grow[1]

In Chapter 1, we looked at Ethernet as a *service* made available by the Media Access Control (MAC) to a MAC client, over an abstract interface called the "MAC Service" interface. In this chapter, we will travel a few steps down the functional stack and examine the interactions at the interface between the MAC and the physical layer.

The MAC interacts with the physical layer through the *ReceiveBit* function, the procedures *TransmitBit* and *Wait*, and the variables *collisionDetect*, *carrierSense*, *receiveDataValid*, and *transmitting*, as specified in Clause 4 of IEEE Std. 802.3. This may look very comforting to programmers and software engineers, but it is too abstract to allow a "MAC component" and a "physical layer component" to interact with each other in an interoperable way at a hardware level.

To allow for interchangeable physical layer components to be built, additional interfaces have been specified just below the abstract interface between the MAC and the physical layer. These interfaces, the Attachment Unit Interface (AUI)[2] and the Media Independent Interface (MII) family, are defined as *hardware* interfaces, specifying

---

[1] Bob Grow, current Chair of the IEEE 802.3 Working Group, summarizing the strength of Ethernet in the March 2005 issue of The Institute.)
[2] The AUI is mentioned for historical reasons only.

the exact number of signal pins, the precise electrical characteristics of the signals, and the connectors to be used.

As the AUI and MII reside below the abstract interface between the MAC and the physical layer, they are actually *part of the physical layer*. The sublayer between the hardware interface and the MAC—the Physical Layer Signaling (PLS) sublayer in the case of the AUI, the Reconciliation Sublayer (RS) in the case of the MII—takes care of the conversion between the electrical signals and the abstract function, procedures, and variables mentioned before. The part below the AUI is called the Medium Attachment Unit (MAU), and the part below the MII is called the physical layer device (PHY).

In what follows, the MAC-PHY rate matching problem at the MII will be studied in detail. It is important to remember that the MII is a hardware representation of the interaction between the MAC and the physical layer, and that the general methodology presented here holds true for systems that do not have an exposed MII, assuming that the same transactions take place at the level of the abstract interface in that case.

Introduction to the Rate Matching Problem

Rate matching is a relatively recent problem in Ethernet. The early PHYs were specified along with a MAC that was capable of running at the exact same bit rate. The MII, the optional interface between the 10 Mbps or 100 Mbps MAC and associated PHY, locks the clocks of MAC and PHY together by means of the TX_CLK and RX_CLK signals that run from the PHY to the MAC. The Gigabit Ethernet and 10 Gigabit Ethernet PHY generations follow the same principle, using the GMII and XGMII interfaces, respectively.

Ethernet owes it popularity to the ability to think of backward compatibility whenever additions or changes to the Ethernet standard are proposed. New PHYs, even those not running at speeds that are an integer power of 10 in Mbps, are expected and must be designed to work with the existing MAC. Unfortunately, the MAC and MII do not provide a straightforward way to do this. The problem first surfaced in 10 Gigabit Ethernet (see IEEE Std. 802.3ae), where it was solved by means of inter-frame spacing (IFS) stretching (see the section "Mac-Centric Solution").

The Ethernet in the First Mile (EFM) Project had its own rate matching issues to deal with, caused by the use of variable-rate Physical Medium Dependents (PMDs) with the existing fixed-rate

MAC. The first occurrence of this problem is in EFM Copper, where DSL-based PMDs offer bitrates that depend on the quality of the loop and the noise and crosstalk conditions. The second occurrence of this problem is in the optical part of the project, where an optional Forward Error Correction (FEC) mechanism (see the section "FEC" in Chapter 7) introduces overhead bytes that need to be transmitted along with the payload bytes.

Operation of MAC and MII

The MII was introduced in Fast Ethernet (IEEE 802.3u) to provide a generic optional interface that would connect any 10 Mbps or 100 Mbps PHY to the Ethernet MAC (see Figure 2.1). The AUI that had been used as a media-independent connector in the 10 Mbps Ethernet generation was deemed unsuitable for Fast Ethernet applications due to its lack of support for management and its CMOS-unfriendly electrical characteristics [21].

The MII is specified as an interface with a data width of 4 bits in each direction, running at a 2.5 MHz or 25 MHz clock, either in half

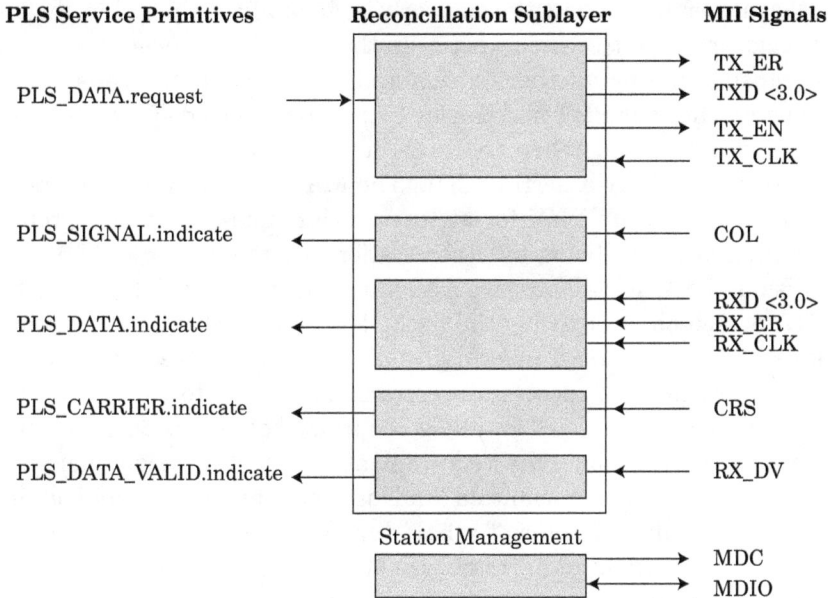

PLS Service Primitives	Reconcillation Sublayer	MII Signals
		TX_ER
PLS_DATA.request		TXD <3.0>
		TX_EN
		TX_CLK
PLS_SIGNAL.indicate		COL
PLS_DATA.indicate		RXD <3.0> RX_ER RX_CLK
PLS_CARRIER.indicate		CRS
PLS_DATA_VALID.indicate		RX_DV
	Station Management	MDC MDIO

Figure 2.1   The Media Independent Interface.[3]

---

[3] From IEEE Std. 802.3-2002, "CSMA/CD," by IEEE. © 2002 IEEE. All rights reserved.

duplex or in full duplex mode. The remaining part of the interface provides all the necessary control signals to allow the MAC and the PHY to interact as required by the Carrier Sense Multiple Access with Collision Detection (CSMA/CD) MAC method. The Reconciliation Sublayer (RS) provides a mapping from the abstract primitives and variables in the MAC specification to the electrical signals on the MII.

In half-duplex mode, Carrier Sense (CRS) is asserted by the PHY when it receives a frame from the MAC over the MII, or when it detects that another end station is transmitting data on the medium. Transmission of a MAC frame from the MAC to the PHY can only be started when CRS is de-asserted. CRS thus ensures that multiple end stations can access a shared medium without interference—barring occurrences of simultaneous transmission, which are detected and signalled to the MAC as "collisions" (COL).

At the start of the transmission of a frame, the Transmit Enable (TX_EN) signal is asserted by the MAC, and in normal operation, it remains asserted until the entire frame has been transmitted at a rate of 4 bits (1 nibble) per clock cycle. When the PHY detects the assertion of TX_EN, it will assert CRS, and keep it asserted until TX_EN is de-asserted.

No matter how small the network is, it is always possible that a MAC attempts to transmit a frame in the brief time between the start of transmission by another station and the moment the transmitted frame reaches our PHY and causes CRS to be asserted. This situation is called a collision. When the PHY detects a collision, it will signal this to the MAC by asserting COL. The MAC will continue sending a data pattern to the PHY for at least 32 bit times, to ensure proper detection of the collision by other stations on the network. After the collision, the MAC will observe a certain random delay (determined by a "truncated binary exponential backoff" algorithm) before attempting retransmission of the frame, to give all stations involved in the collision a fair chance of successfully retransmitting. Up to 16 retransmission attempts are allowed for any given frame before it is declared lost.

In half-duplex mode, the requirement to detect collisions before a frame is completely transmitted (and lost), puts an upper bound on the collision domain diameter of a half-duplex network. For a 100 Mbps network with a minimum frame size of 64 bytes, the maximum diameter is 200 m. To avoid having to shrink the network to 20 m at 1 Gbps, the Gigabit Ethernet standard requires carrier extension at the end of short frames, to ensure a minimum duration of at least 4096 bit times for every transmission. It is allowed to group several consecutive small frames into a single carrier event (frame bursting) to improve the efficiency of the transmissions. At 10 Gbps, even such tricks could not have resulted in a sufficiently large and efficient half-duplex network, so half-duplex mode is forbidden at speeds above 1 Gbps.

In the receive direction, the transmission of data from the PHY to the MAC is signalled by the assertion of Receive Data Valid (RX_DV). In normal operation, it remains asserted until the entire frame has been transmitted at a rate of 4 bits per clock cycle.

This basically sums up the operation of the CSMA/CD protocol, which is at the heart of Ethernet. Note that the specification of half-duplex operation does not prohibit the simultaneous transmission and reception of data across the MII.

In full-duplex mode, TX_EN and RX_DV are used in the same way as in half-duplex mode. The use of the CRS and COL signals is not specified in full-duplex mode.

The TX_ER and RX_ER signals are used by the MAC and the PHY, respectively, to signal an error in the frame that is being transmitted over the MII.

As shown, MAC/MII combinations work exclusively at data rates of 10 Mbps or 100 Mbps (functionally equivalent interfaces exist at 1 Gbps and 10 Gbps, respectively, called GMII and XGMII). The xDSL PHYs to be used for EFM over copper operate at rates determined by the physics of the loop and the configuration by the operator. The xDSL data rate will rarely be exactly 10 Mbps or 100 Mbps, and even then it is susceptible to time-dependent variations. This explains the need for MAC-PHY rate matching.

MAC-Centric Solution

In the course of the 10 Gigabit Ethernet project, a 10GBASE-W PHY was specified that was to be compatible with 10 Gigabit WAN PMDs as used in SONET/SDH networks (specifically, for the SONET STS-192c and the SDH VC-4-64c formats). Such "10 Gigabit WAN PMDs" actually run at a rate of 9.58464 Gbps. To match the rate of the 10 Gigabit Ethernet MAC to the 9.58464 Gbps PMD, the 10GBASE-R Physical Coding Sublayer (PCS) is allowed to insert additional idle characters in the receive path and delete idle characters belonging to a "stretched" IFS in the transmit path. IFS stretching is performed by the MAC, which keeps track of the length of the frame to be transmitted, and extends the IFS with a number of bits that is proportional to that length.

For IFS stretching to work correctly, the MAC needs to "know" the bitrate ratio between the MAC and the PHY. This is not a problem if the ratio is fixed, as is the case for 10GBASE-W, but in other cases, some method must be provided to convey bitrate information from the PHY to the MAC. This would complicate matters in implementations that have a physical interface between MAC and PHY (the bitrate information would have to be passed from the PHY to the MAC over that interface), which is why this method was not selected for EFM. In an integrated MAC+PHY implementation, in which management information can be exchanged more freely between MAC and PHY, the use of IFS stretching is of course allowed without loss of interoperability.

PHY-Centric Solution

The PHY-centric rate matching solutions rely on triggers from the PHY to the MAC to control the bitrate between MAC and PHY. A good understanding of the MAC and of the MII is required to analyze these solutions, especially the one rate matching mechanism selected by the EFM Task Force, known as "CRS deference."

Sub-Frame Rate Matching versus Frame-Based Rate Matching

In sub-frame MAC-PHY rate matching, transmission of data from the MAC to the PHY can be suspended (or slowed down) at any time.

A first way to accomplish this is to add a suspend signal to the interface, which can be asserted by the PHY at any time, causing the MAC to stop transmitting any further nibbles to the PHY until the signal is de-asserted. An interface with such a signal is known as a variable bit rate MII (VMII), because the suspend signal can be used to pull data over the interface at any desired rate (limited by the clock speed). A similar signal is used to push received data to the MAC at a pace enforced by the PHY. Existing implementations of the VMII use combinations of existing signals to suspend transmission rather than adding more pins to the MII.

A second way to implement sub-frame MAC-PHY rate matching uses the clock signals TX_CLK and RX_CLK. The IEEE 802.3 MAC is specified in a way that is nearly completely independent of the clock rate. If the PHY generates clock signals that correspond to the data rates on the line, the MAC should be able to process data at that rate. Some existing implementations of MACs may depend on the standardized clock rates to calculate certain real-time intervals, and these MACs may therefore dysfunction under clock stretching.

Neither of these sub-frame rate matching methods is compliant with the existing specifications in IEEE Std. 802.3. Therefore, the preference of the EFM Task Force went to frame-based rate matching.

For frame-based MAC-PHY rate matching, the rate of the MII must be greater than or equal to the PHY rate. In receive direction, a complete MAC frame is received from the remote end at line rate, buffered, and transmitted to the MAC Sublayer at MII rate. In transmit direction, a complete MAC frame is received from the MAC Sublayer at MII rate, buffered, and transmitted to the remote end at line rate. Since the line rate is less than the MII rate, the PHY must be able to keep the MAC from transmitting any more frames when the PHY's transmit buffer is full.

From a standardization point of view, the MAC-PHY rate matching problem had to be solved in a way that required the least possible changes to existing specifications and the highest possible probability of backward compatibility with existing implementations. This ruled out any changes to the specification of the MII, such as VMII and clock stretching. A frame-based MAC-PHY rate matching mechanism was preferred, which led to the adoption of the "CRS Deference Method."

The CRS Deference Method

As mentioned earlier, the CRS signal was originally defined in half-duplex mode only. The operation of the CRS deference method with an existing pre-EFM MAC requires the MAC to be configured in half-duplex mode. Because transmit and receive functions of the MAC are defined completely independently, however, the half-duplex mode does not actually prevent transmission and reception from happening simultaneously in a standard-compliant MAC [32].[4]

In the receive direction, data coming in from the subscriber line at line rate is stored in a buffer residing between the γ-interface and the MII. Whenever the reception of a MAC frame is completed, it is sent to the MAC at MII rate in the usual way (by asserting RX_DV, but not CRS, for the duration of the transmission).

In transmit direction, data are sent from the MAC to the PHY at MII rate. The PHY asserts CRS shortly after the assertion of TX_EN by the

---

[4] Special provisions have been taken in the specification of MAC-PHY Rate Matching to ensure that true half-duplex operation can also be used, for MACs that are not capable of transmitting and receiving simultaneously in half-duplex mode. This special mode is also useful for PHYs, which use an alternative interface to the MAC, such as SMII or RMII.

MAC, and keeps it asserted for the duration of the transmission of a frame. In the PHY, a buffer residing between the MII and the γ-interface collects data for encapsulation. If, at the end of MAC frame transmission, the buffer does not have sufficient space available to handle another incoming MAC frame, the PHY shall keep CRS enabled until it is ready for another frame. The CRS signal has the effect of deferring the transmission of the next frame.

It is important to note that the size of the data buffers in the PHY required for the CRS deference method depends on the maximum data frame size (currently 1522 bytes). As this size is subject to change in the future, to accommodate tags for new services or protocols, a future-proof implementation should be dimensioned for substantially larger frames.

The Simplified Full-Duplex MAC

Near the end of the EFM project, Annex 4A was created, containing a slimmed-down, full-duplex only version of the IEEE 802.3 MAC. Strictly speaking, it is *not* a new MAC; it is simply a reiteration of the old MAC with the legacy half-duplex requirements left out. However, the behavior with respect to the *carrierSense* variable is new. Hence, the title of this chapter.

A special characteristic of the Annex 4A MAC is that it is capable of deferring transmission of a frame based on the value of the boolean variable *carrierSense*—this is essentially the half-duplex behavior explained earlier. To enable *carrierSense* reactivity, the variable *carrierSenseMode* must be set to TRUE. If *carrierSenseMode* is FALSE, the *carrierSense* is ignored, and another means of rate adaptation must be used (e.g., relying on higher layers to stretch the IFS appropriately).

Note that the specification of the MII leaves CRS undefined when the MAC is in full-duplex mode. For *carrierSense* to be properly observed in "Annex 4A mode," the value of the CRS signal must be mapped to *carrierSense* even though the MAC is operating in full-duplex mode. This is a new requirement introduced by EFM. When using existing MACs that don't observe CRS when configured in full-duplex mode, half-duplex mode must be used for deference.

Rate Matching with RMII and SMII

The high pin count of the MII makes this interface inappropriate for use in multi-port devices such as Ethernet switches. For this reason, vendors have come up with alternative interfaces for the interconnection of MACs and PHYs. Both alternative interfaces, discussed next, use CRS differently from the original MII (the signal only reacts to received data), which necessitates changes to the CRS deference mechanism. The EFM Copper PHYs are capable of adapting their behavior to comply with the needs of these interfaces.

RMII. The Reduced MII interface is a proprietary interface promoted by AMD, Broadcom, National Semiconductor, and Texas Instruments [2]. It operates at a clock speed of 50 MHz, transmitting di-bits in each direction; this saves four pins on the data path only with respect to the MII. Four more pins are saved by collapsing CRS and RX_DV into a single CRS_DV signal, by omitting TX_ER and COL, and by using a single reference clock for transmit, receive, and control.

The fact that CRS and RX_DV appear as a single CRS_DV signal complicates the CRS deference method in systems that use the RMII. The CRS_DV signal is asserted when valid received data are being presented to the MAC. In half-duplex mode, the MAC defers any transmissions to the PHY as long as CRS_DV is asserted. This means that simultaneous transmission and reception is impossible in this case (for EFM Copper PHYs that support the Clause 45 management interface, the *MII receive during transmit* register must be cleared to *binary 0* in this case).

SMII. The Serial-MII interface is a proprietary interface promoted by Cisco Systems [3]. It operates at 125 MHz, serially transmitting 10-bit words, containing one byte of information and two bits representing CRS and RX_DV (in the receive path) or TX_EN and TX_ER (in the transmit path). Using a single 125-MHz reference clock (CLOCK) and a single-word synchronization signal (SYNC) for information transfer in both directions, the total pin count can be as low as four. A system needs only two additional pins for every added port.

The SMII may optionally be source synchronous. In that case, the SYNC signal is replaced by RX_CLK, and RX_SYNC signals for the receive direction, plus TX_CLK and TX_SYNC signals for the transmit direction.

Collisions occur when SMII signals CRS and TX_EN are simultaneously asserted. This means that CRS must never be asserted for rate matching purposes when a transmit frame is being transferred over the SMII (for EFM Copper PHYs that support the Clause 45 management interface, the *TX_EN and CRS infer a collision* register must be set to *binary 1* in this case).

A Note on Flow Control

In full-duplex mode, the use of the MII signal COL is not defined; because the transmit and receive channels are independent, collisions cannot occur. A specific flow control mechanism was added to the standard by project 802.3x, to allow two stations on a full-duplex link to signal congestion to each other, for example, when the receive buffer of one of the hosts is (nearly) full. This mechanism relies on PAUSE frames.

PAUSE frames are generated in the optional MAC Control sublayer, a functional layer between the MAC and the MAC client (or more correctly, the MAC control client). The PAUSE frames are sent to the dedicated global multicast address 01-80-C2-00-00-01 and contain a PAUSE opcode and a pause_time operand in the range between 0 and 65535 pause_quanta (pause_quanta being equal to 512 bit times). PAUSE frames are received by the MAC Control sublayer of the system on the other end of the link, if present, and result in a temporary suspension of all transmissions from that system.[5]

It has been proposed to use PAUSE frames for the purpose of MAC-PHY rate matching in EFM. This would involve PAUSE frames generated in the physical layer to be passed up the MII toward the MAC control layer. This abuse of the PAUSE mechanism is not only architecturally impure (MAC control frames ought to be exchanged between peer MAC control layers, not between different layers at a single end of the link), and it would also render the PAUSE mechanism useless for its original purpose, which is signaling congestion at the receiving end.

Despite these arguments against the use of PAUSE frames for MAC-PHY rate matching, this mechanism is actually known to work quite well, and has been used in the design described in Annex B. Especially in cases where one is stuck with a MAC implementation that does not have proper CRS functionality, PAUSE frames can be the easy way out.

At the time of writing, a task force on "Congestion Management" is studying possible extensions of the PAUSE frame mechanism to include more sophisticated congestion indications.

---

[5] Note that on a full-duplex LAN, there are always exactly two stations (one of which may be a port of a bridge), so there can be no confusion as to which station receives the PAUSE frame. PAUSE frames must not be forwarded to other LANs by MAC bridges.

3

EFM over Copper
General Architecture

"DSL will be around for the next 100 years..." —JOHN CIOFFI

When discussing Ethernet in the First Mile's (EFM's) point-to-point copper solutions, there are two misconceptions that I encounter very frequently, representing two opposite and incorrect views of the Task Force's accomplishments. Before starting the detailed discussion of the EFM copper architecture in this chapter, and the 10PASS-TS and 2BASE-TL specifics in the next two chapters, let us get these misconceptions out of the way.

The first view is that the EFM copper Physical Layer entity sublayers (PHYs) 2BASE-TL and 10PASS-TS are limited to bitrates of 2 Mbps and 10 Mbps, respectively. This false idea probably results from mistaking the Task Force's *objectives* (specifically the last two bullets of objective 2 in Annex A) for the *standard.* The objectives were set to guide the work of the Task Force, and if the port types that were eventually standardized meet these objectives, they are by no means limited by them.

The second view is that the EFM copper PHY specifications are identical to the existing Single-Pair High-Speed Digital Subscriber Line (SHDSL) and Very-High-Speed Digital Subscriber Line (VDSL) standards "with Ethernet on top." This view sticks to the Task Force's *baseline proposals* too closely, and ignores the fact that the Task Force has actually done a lot of work on top of what already existed in DSL standards. Not only are the frame encapsulation and bonding methods used by EFM original, but there are a number of changes and restrictions with respect to the referenced standards in the lower parts of the physical layer that make 2BASE-TL and 10PASS-TS different PHYs in their own right. The Task Force was nevertheless careful not to introduce changes that would preclude the implementation of "multimode" components or network nodes, which can operate in an EFM mode and in a traditional DSL mode.

Point-to-Point Copper Objectives

The earliest Call-for-Interest presentations on "Ethernet in the Last Mile" (such as the one by Howard Frazier [22], later to become Chair of the EFM Task Force), identified the ELM medium as "UTP POTS" and the PHY as something very much like VDSL. Nevertheless, the Study Group did not approve a single copper-oriented goal at its first meeting in January 2001, partly due to the presence of a great number of people seeking to standardize fiber-based solutions, and partly because of the lack of consensus among the copper-oriented attendees. By the next meeting, the copper/fiber ratio of the audience had readjusted to a level where it was clear that "point-to-point copper" was going to be part of the EFM project, and the debate on the detailed objectives could start.

Short-Reach Objective

10PASS-TS was specified as a solution for the first *concrete* objective regarding operation over point-to-point copper that the EFM Task Force adopted: a *"PHY for a single pair of voice-grade copper, distance ≥750 m, speed ≥10 Mbps Full Duplex."*[1]

The most important difference with existing twisted pair Ethernet standards is that the EFM Task Force aimed at specifying a PHY operating over a single pair, as opposed to the two pairs required for 10BASE-T, 100BASE-TX, and 100BASE-T2 or the four pairs required for 100BASE-T4, 1000BASE-T, and 10GBASE-T.[2] Although the group did review a proposal based on a lower speed, longer range version of 1000BASE-T at an early stage of its work, this proposal became irrelevant after the adoption of the quoted objective.

---

[1] In the form in which this objective was first adopted, it specified a distance of 2,500 feet and a speed of 10 Mbps aggregate, but these requirements were later considered too loose.

[2] 10GBASE-T is still under definition at the time of writing.

A second difference is the qualification of the wire: In EFM, the wire is supposed to be voice-grade (i.e., category 3 or worse). Previous twisted pair Ethernet standards specify the need for category 3 (10BASE-T, 100BASE-T4, 100BASE-T2), category 5 (100BASE-TX), or category 5e (1000BASE-T), and a maximum distance of 100 m, assuming a single, straight stretch of cable (an overview is given in Table 3.1). The environment in which EFM is deployed does not allow for any tighter qualification than "voice-grade," and it is well known that the actual cable may contain impairments such as gauge changes and bridged taps.

The short-reach objective had "VDSL" written all over it. VDSL is the fastest xDSL flavor currently available, and several pre-standard VDSL implementations exist that use Ethernet rather than Asynchronous Transfer Mode (ATM) transport (the brand names of these products tend to refer to their Ethernet nature). Thanks to its high bitrate, the reasons for making lower-speed xDSL flavor ATM-based do not necessarily apply to VDSL; Quality of Service (QoS) and queueing issues are less of a problem the more bandwidth is available.

The trouble with VDSL is that several other Standards Development Organizations (SDOs) had been trying in vain to select a single modulation ("linecode") for this xDSL flavor. Both the proponents of Discrete Multi-Tone (DMT) modulation and the proponents of Quadrature Amplitude Modulation (QAM) were well represented in the EFM Task Force. The stage was set to start fighting the linecode war on yet another front....

Long-Reach Objective

The EFM Task Force and its Copper Sub Task Force spent a lot of time debating the need for a long-reach objective. When the Task Force first agreed to have a long-reach objective, a year after the original EFM Call for Interest, they immediately had two: *"PHY for single pair non-loaded voice grade copper, distance ≥3700 m, 0.5 mm, [bitrate] ≥4Mbps"* and *"PHY for single pair non-loaded voice grade copper, distance ≥4600 m, 0.4 mm, [bitrate] ≥256 kbps,"* plus the objective to specify an optional mode of operation using multiple copper pairs. The 4 Mbps objective was approved by the Task Force with 68 in favor and 4 against, while the 256 kbps objective just barely reached the 75 percent threshold with 62 "yeas" and 18 "nays."

The IEEE 802.3 "CSMA/CD" Working Group, the parent organization of the EFM Task Force, was not happy with the long-reach objectives. Especially the 256 kbps objective appeared to be unacceptable: Not only is it outside the formally approved scope for the EFM project (*"... operating speeds within the scope of the current IEEE Std. 802.3 and approved new projects,"* which places the EFM target rates between 1 Mbps and 10 Gbps), but it would also be *"a significant change to what Ethernet has delivered traditionally and could be interpreted as just an attempt to cash in on the Ethernet name,"* as reflected by one of the opinions in the minutes of the Working Group closing plenary in November 2001.

An effort by the Chair to reword the long-reach objectives and to resubmit them to Task Force did not improve things. By the end of the next Working Group plenary meeting, in March 2002, the Task Force had no long-reach objective to offer at all and had missed its project milestone of approving baseline proposals for all objectives. Interestingly, at this meeting, a majority voted in favor of a motion to give up the goal of specifying any copper PHYs and simply specify a generic Ethernet-over-xDSL Adaptation Layer instead (as described in the section "From PTM-TC to EFM-TC").

Members of the Copper Sub Task Force continued to work behind the scenes on a longer-reach objective, with strong support from participating operators. This initiative was welcomed by the Task Force leadership, which hoped that a solution to the long-reach problem would help break the short-reach deadlock, which had formed in its wake.[3]

At the Vancouver plenary session in July 2003, the Working Group ratified the Task Force's new and final long-reach objective: *"PHY for single pair non-loaded voice grade copper distance ≥2700 m and speed ≥2 Mbps full duplex."* Even the most casual observer could see that this agreement had been built *between* meetings; before the meeting's presentation submission deadline—that is, before the motion to adopt the long-reach objective had even been made—presentations were submitted in support of the two candidate PHYs for the long reach: ADSL2 "Annex J" and SHDSL. Thus started a second linecode war, this time for the long reach, which would take six more months and another trip to Vancouver to be resolved in favor of SHDSL (see Chapter 5).

---

[3] Approval of the baseline proposal for the long-reach objective was apparently being blocked by two main groups: members who felt that EFM over point-to-point copper made no economic sense if it was specified on short loops only, and people who felt that approving a VDSL-based baseline was useless if there was no agreement on how to select a linecode.

TABLE 3.1   Overview of Twisted Pair Port Types

Port Type	No. of Pairs	Cable Qualification	Reach
2BASE-TL	1	Voice-grade	2700 m
10PASS-TS	1	Voice-grade	750 m
10BASE-T	2	CAT 3	100 m
100BASE-TX	2	CAT 5	100 m
100BASE-T4	4	CAT 3	100 m
100BASE-T2	2	CAT 3	100 m
1000BASE-T	4	CAT 5e	100 m
10GBASE-T	4	CAT 6/7	55 m–100 m

Port Type Names

Naming the new EFM Copper port types under development was not
a trivial exercise. To stick with traditions and maximize the transparency of the names, it was deemed necessary to build up the names
from existing elements.

Unlike nearly all existing Ethernet PHYs, the VDSL-based short-
reach PHY is a *passband* technology. The keyword "BASE," as used in
all the other twisted-pair PHYs listed in Table 3.1, would therefore be
inappropriate as a naming basis. The only previously adopted passband PHY is 10BROAD36, but it was the feeling of the Task Force that
reusing the keyword "BROAD" would bring back memories of a less
than successful technology. "PASS" was a generally acceptable alternative.

With an objective of delivering 10 Mbps over a single pair of voice-
grade copper, 10PASS-T1 would seem to be a logical result. "T1"
sounds like the name of a 1.5 Mbps PDH link, however, and the name
of another standardization body! "VG" for "voice-grade" sounded polit-
ically incorrect to anyone who remembered the VG-AnyLAN war, a
time when division over a "100BASE-VG" proposal caused a number of
members of the "Fast Ethernet" Task Force to go off and form Working
Group 802.12, which created the "Demand Priority" standard, now
withdrawn as a standard and practically extinct as a technology. As a
result, 10PASS-T was adopted for the time being, with a lingering dis-
comfort over the fact that this "T" was quite a different medium than
the "T" in the existing 10BASE-T specification.

The short-reach PHY had its own problems. Although SHDSL is a
baseband technology, making the use of the keyword "BASE" obvious,

the name 2BASE-T was allegedly already taken—it seems to be a long forgotten proprietary technology transporting 2 Mbps over 50 Ω coaxial cable.[4] For this reason an "L" (for long-reach) was added to the name, resulting in "2BASE-TL." Logically, the ADSL-based rival proposal for the long-reach objective was named "2PASS-TL." For consistency, "10PASS-T" was renamed "10PASS-TS" ("S" for short-reach).

With PHYs based on xDSL transceivers, another naming issue had to be resolved: how to distinguish the two different ends of the link. The xDSL transceivers used at the central office (CO) side of a link are physically different from the transceivers used as customer premises equipment (CPE)—not only are the transmitted electrical signals different, but the role played by each side of the link in initialization and management procedures is also fundamentally different. Keeping with the tradition of VDSL, where "VTU-O" and "VTU-R," respectively designate the VDSL Transceiver Unit at the CO side (Optical) and the CPE side (Remote), EFM Copper introduced "10PASS-TS-O" and "10PASS-TS-R." The same convention applied to 2BASE-TL led to "2BASE-TL-O" and "2BASE-TL-R," which may look a bit unusual to readers familiar with the SHDSL standards, in which "STU-C" and "STU-R" are used.

The First Mile of Copper

Limitations of the Local Loop

The challenge of making a physical layer specification for the subscriber's telephone line is in the transmission characteristics of the twisted pair. Subscriber lines often have a length of several kilometers and are composed of different pieces of wire with different gauges. Some lines have one or more *bridged taps*; these are pairs that are connected in parallel to the subscriber line (often put in place to allow the same line from the central office to be connected to any one out of a certain set of subscribers). Finally, subscriber lines run in bundles containing tens to hundreds of individual pairs, all of which will cause some electromagnetic signal leakage (*crosstalk*) from one pair to another.

Transfer function.   The transfer function $H(f)$ of a channel describes the relation of the signal coming out of the channel to the signal going into

---

[4] It is difficult to find an authorative reference on 2BASE-T (or "2BaseT"). Several definitions can be found on the Web, some identifying it with thin Ethernet, some with 2 Mbps over coax, and some with 8 Mbps over coax.

the channel as a function of frequency. For a copper pair, the main trend is that attenuation increases with frequency and with distance. This means that the useful bandwidth of a copper pair (determined by the highest frequency at which a useful signal can be made out between noise and crosstalk for a given transmitted power spectral density) decreases with increasing loop length. Gauge changes and bridged taps cause reflections that show up as (series of) dips in the transfer function, further reducing the information capacity of the channel.

Crosstalk.   Two main categories of crosstalk are distinguished: near-end crosstalk (*NEXT*) and far-end crosstalk (*FEXT*). NEXT is signals from other transmission equipment at the same end of the cable bundle, leaking into the receiver. This occurs typically at the CO, where large numbers of transmitters are colocated. FEXT is generated by transmission equipment at the other end of the cable bundle. Although FEXT gets attenuated by the copper pair before it reaches the receiver, it can be more damaging to the system because in the case of self-FEXT,[5] it occupies the exact same spectrum as the useful signal. Moreover, the FEXT source may be closer to the receiver than the target transmitter. Different analytical models of NEXT and FEXT power coupling functions exist for use in simulations (e.g., those specified in American National Standard T1.417-2001):

$$\text{NEXT}(f,n) = S(f) \cdot X_N \cdot n^{0.6} \cdot f^{\frac{3}{2}} \tag{3.1}$$

$$\text{FEXT}(f,n,l) = S(f) \cdot |H(f)|^2 \cdot X_F \cdot n^{0.6} \cdot l \cdot f^2 \tag{3.2}$$

In these equations, $S(f)$ is the power spectral density (PSD) of the disturber, and X_N and X_F are wire-related constants. As can be seen from the models, both NEXT and FEXT increase less than linearly with the number of disturbers n, and more than linearly with frequency f. The factor l (loop length) in the FEXT expression models the fact that the longer the distance over which the disturber and the victim pair run through the cable together, the more crosstalk will couple into the victim pair. It is important to note that actual crosstalk coupling functions will never be as smooth as the models; also, the crosstalk coupling functions for any two pairs out of a cable will be different.

---

[5] This is FEXT generated by a system that is identical to the victim system.

Radio frequency interference (RFI).   All electronic circuits send a certain amount of energy into space under the form of electromagnetic waves. When these waves are in the radio frequency range, roughly the spectrum from a couple of kHz up to several GHz, they easily bounce off or diffract around obstacles and find their way into other electronic circuits—with effects that are sometimes desired (as in radio receivers), but most often undesired. In xDSL systems, the unwanted interference will typically render certain narrow frequency bands useless for information transmission (e.g., bands around the carrier frequency of AM radio stations). Conversely, certain frequency bands must not be used by xDSL transmitters, to avoid harmful interference towards HAM radio equipment.

Noise.   All other nonsignal energy that is detected by the receiver will be labeled *noise*. For modeling and simulation purposes, a floor of white gaussian noise is often assumed at −140 dBm/Hz. In addition to this constant noise floor, there may be a certain level of impulse noise, occurring for a short period of time at irregular intervals.

The influence of channel transfer function, crosstalk, and noise on channel capacity over a given bandwidth W is given by the Shannon-Hartley capacity theorem [38]:

$$C = W \cdot \log_2(1 + \mathrm{SNR}) \qquad (3.3)$$

In this theorem, the main variable governing the capacity of a channel is the signal-to-noise ratio (SNR). In the classical form, shown here as Equation 3.3, the noise is assumed to be additive white gaussian noise (AWGN). If this assumption does not hold, the frequency-dependence of the noise must be taken into account, resulting in Equation 3.4. Noise includes all forms of nonsignal energy, including crosstalk and RFI. Figure 3.1 illustrates the theorem.

$$C = \int_0^W \log_2 [(1 + \mathrm{SNR}(f)] \, df \qquad (3.4)$$

The Shannon-Hartley capacity theorem indicates the maximum bitrate that can be achieved at arbitrarily low bit error ratio on a channel with a given SNR. The theorem doesn't tell us *how* to achieve this bitrate. The fact that SNR is a statistic variable, calculated using a time-averaged noise quantity, which is assumed Gaussian, implies that different time intervals may have different noise characteristics. To benefit from the "average capacity," the transmitted information should

$$C = \int_{f_{min}}^{f_{max}} \log_2(1 + SNR(f))df$$

PSD
[dB]

Received Signal [dBm/Hz]

SNR [dB]

Received Noise [dBm/Hz]

Frequency [Hz]

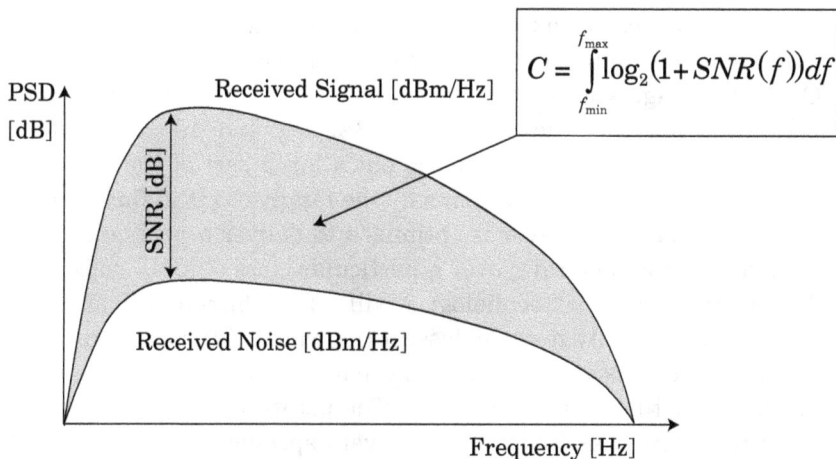

Figure 3.1   The Shannon-Hartley channel capacity theorem. (*Source: EFM Task Force* *[4]*)

be spread in time; this is achieved by choosing low symbol rates, coding techniques, and interleaving (the use of these techniques is discussed in the chapters dealing with the different PHY specifications). The disadvantage of spreading information in time, is the extra latency that is introduced between transmission and reception of the data.

A practical indication of the capacity of a communication system is obtained when the efficiency of the modulation technique is taken into account; this is achieved by dividing the SNR in Equation 3.4 by a modulation-specific "SNR gap" (for QAM modulation at a bit error ratio of 10^{-7}, a gap of 9.75 dB is typically assumed). The formula can be further modified to take the effect of forward error correction into account (a multiplicator designated as "coding gain"). Finally, an "SNR margin" of 5 or 6 dB is typically employed; the actual bitrate of the system is chosen as if the available SNR were lower than the actual one. When operating in this manner, the noise on the channel may (temporarily) increase by up to as much as the SNR margin, without risk of exceeding the targeted bit error ratio (BER) (this implies that the average bit error ratio at nominal noise levels is much lower than the target).

Although previous Ethernet standards have had to cope with the same media impairments as the EFM Copper standard, the general practice in IEEE 802.3 has been to specify the media along with the PHY. Thus, 10BASE-T transceivers have been engineered and specified to operate over two pairs of category-3 wiring, at distances of up to 100 m, and 100BASE-TX will run over two pairs of category-5 wiring at distances of up to 100 m. If wires of lower quality are used, or if the distance limitation is violated, the system will no longer work.

Given the huge variety in the quality of subscriber loops, a tight specification such as the ones for 10BASE-T and 100BASE-TX is impossible to achieve without ruling out a large part of the potential market. This is part of the reason why the family of xDSL flavors is so extensive. Each xDSL flavor is optimized to deliver a particular type of service (not just one rate) over a particular class of local loops.

A statement such as "technology x will offer a bitrate of y at a distance of z" is virtually meaningless in the world of twisted-pair communications, unless the specific conditions of the transmission (e.g., wire gauge, disturbers) are specified. The point-to-point copper objectives of the EFM Task Force were not very specific in that respect. It is fair to say that the two copper PHYs specified by EFM meet their respective objectives *under certain conditions*—as did a number of competing proposals; the actual performance varies with the circumstances of the deployment (or the parameters of the simulation), so the Task Force went through dozens of rate-reach simulation plots and tried to select the best tradeoff.

Spectrum Management

Unlike PON-, radio-, or cable-based access networks, the local telephone loop as used by xDSL technologies is not usually thought of as a shared medium. Although xDSL is certainly a point-to-point connection at the data-link level, telephone wires are grouped in cables containing anywhere between one and several hundreds of pairs, which undergo electromagnetic coupling. As we mentioned earlier, the signals on these pairs will suffer from crosstalk from adjacent pairs, which causes severe deterioration of system performance. Fixed-rate systems need a certain modulation-dependent minimum SNR to operate at a preselected BER. Rate adaptive systems may lower their bitrate in order to preserve BER at the desired level. Preserving minimum bit rate and BER characteristics in the presence of a variety of xDSL systems is the goal of spectrum management.

One dimension of the problem of spectrum management is duplexing. The method used to transmit bidirectionally over a single twisted pair, impacts the self-crosstalk and echo experienced by the system.

Frequency domain duplexing (FDD).   This duplexing method relies on a spectral separation of upstream and downstream traffic. Data are modulated on two disjoint sets of carriers, with baud rates chosen in such a way that the spectra of the modulated carriers do not overlap. This method eliminates most of the NEXT, as one system's transmit signal does not overlap with another system's receive spectrum. VDSL and ADSL are examples of FDD systems.

Echo canceling (EC).   In echo canceled systems, a certain part of the spectrum is used for simultaneous transmission in both directions. Systems have to remove any traces of their own transmit signal (echo) from the detected receive signal. The frequency range over which echo canceling is useful as a duplexing technique is limited by the NEXT coupling function, because from a certain frequency on, the NEXT will overwhelm the receive signal energy. SHDSL and 1000BASE-T are examples of an echo canceled system. In ADSL, the downstream spectrum may optionally overlap with the upstream spectrum, which results in a partially echo-canceled system.

Time domain duplexing (TDD).   In the case of TDD, the same transmit spectrum is used in both directions in an alternating way. The advantage of this approach is that any asymmetry ratio between upstream and downstream can be obtained simply by adjusting the time allocated to each direction of transmission. NEXT can be eliminated, on the condition that all systems operating in the same cable bundle are synchronized (i.e., they must transmit at the same time and receive at the same time). This imposes the same asymmetry ratio on all subscribers. Moreover, in the recently "unbundled" telecom environment, it would be very hard to implement synchronization across equipment owned by different operators. Early DMT-VDSL systems were based on TDD, as are certain nonstandard subscriber line technologies.

Dual simplex.   Finally, duplex operation can be achieved by combining two simplex channels (each using a different copper pair). T1 lines, HDSL, 10BASE-T, and 100BASE-TX are all examples of transmission systems that use different twisted pairs for each direction of transmission.

The IEEE 802.3 MAC runs either in *half duplex* mode or in *full duplex* mode.[6] In the early shared media systems, such as 10BASE5 and 10BASE2, only one transmitter may send data at a time, or else the signal on the media becomes undecipherable to all receivers. Two

---

[6] Full duplex is the only allowed mode of operation for transmission speeds above 1 Gbps.

point-to-point twisted-pair PHYs (100BASE-T4, 100BASE-T2) use multiple pairs for transmission in one direction and therefore cannot transmit in both directions at the same time. 10BASE-T and 100BASE-TX are point-to-point twisted-pair links, which provide a separate pair in each direction. 1000BASE-T transmits on four pairs, in both directions simultaneously. In these PHYs, half duplex is not a physical limitation of the transmission medium and modulation, but a mode of operation of the MAC that can be selected by the user.

The problem of *spectral compatibility* of different wireline systems is addressed by several *spectrum management* methods and standards. American National Standard T1.417 is the main reference document on this topic.

This spectrum management standard defines spectral compatibility in terms of the minimum SNR margin (or bitrate) that must be met by a set of "basis systems" in the presence of a certain mix of disturbers. It specifies two methods to demonstrate spectral compatibility of new systems with the existing basis systems.

Method A defines a series of spectral classes, specified by their upstream and downstream PSD masks, which are known to be spectrally compatible with the basis systems up to a certain loop length (the deployment guideline). Any system that uses upstream and downstream PSD masks, which are completely contained within those defined for one of the classes, is spectrally compatible up to the deployment guideline of that class.

Method B (described in Annex A of T1.417) is an algorithmic method, in which the SNR margin (or bitrate) of each of the basis systems is calculated under the impact of certain mixes of reference disturbers and the new systems under test. The system under test is declared spectrally compatible if and only if the SNR margin (or bitrate) reached by each of the basis systems in the presence of the disturber mix containing the system under test is greater than or equal to the SNR margin (or bitrate) reached by that basis system in the presence of the reference disturbers. If it occurs that the system passes the test on shorter loops, but not on longer ones, this also results in a deployment guideline.

The methods described here are known as "static spectrum management" methods (coordination level 0). Dynamic spectrum management (DSM) is the capability of a system to dynamically adapt the transmit PSD of each line in order to increase the capacity of the entire loop bundle. DSM comes in different levels of complexity, ranging from autonomous per-line power back-off aiming at crosstalk avoidance (coordination level 1), over coordinated power allocation (coordination level 2), to fully coordinated multi-user transmission aiming at crosstalk mitigation (coordination level 3).

By avoiding or mitigating the effects of crosstalk, DSM can achieve much higher average bitrates (or, equivalently, longer reaches for the same bitrate) over sets of loops that are crosstalk-coupled [10]. The EFM standard does not explicitly address DSM, but the selected PHYs all support some level of control over the transmitted PSD, such that DSM at level 1 or level 2 can be implemented in the management system.

The Ethernet-over-xDSL Model

The ITU-T specifies its xDSL transceiver recommendations according to a layered reference model shown in Figure 3.2.

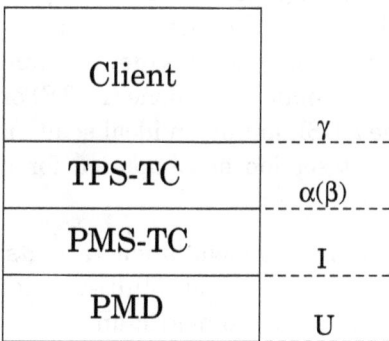

Figure 3.2   ITU-T reference model for xDSL transceivers.

The bottom layer of the reference model is the Physical Medium Dependent (PMD) layer, which contains the flavor-specific physical functions. The PMD transmits and receives a stream of symbols, suitably designed to carry information over the medium with a sufficiently low probability of error. Different xDSL flavors use different modulation techniques and different spectra to provide specific services over specific types of loops in an optimal way. The PMD layer is terminated by the U-interface at the bottom (i.e., the electrical interface with the medium) and by the I-interface at the top (a purely functional interface).

The Physical Medium Specific Transmission Convergence (PMS-TC) layer is the layer above the PMD layer. The PMS-TC transmits and receives information encoded as bits, to be loaded onto the PMD symbols. It has no actual awareness of the physical medium, other than the bitrate that it experiences at the I-interface. It contains functions such as forward error correction, framing, and scrambling, which are designed to work well with a specific modulation technique. Therefore, the PMS-TC is flavor-specific.[7] The PMS-TC is terminated at the top by the functional α-interface (CO side) or β-interface (CPE side).

The PMD and PMS-TC layers have been specified for a growing number of xDSL flavors, each optimized for a certain loop reach or service mix. The following flavors currently have ITU-T Recommendations.

G.991.1–2: HDSL, SHDSL. These flavors were initially intended as a replacement for T1/E1 systems. They use Pulse Amplitude Modulation (PAM), and offer symmetric services (targeted at business users). See Chapter 5 for a detailed discussion of SHDSL.

G.992.1–5: ADSL, G.lite, ADSL2, G.lite.bis, ADSL2+. A growing family of specifications based on the DMT modulation technique. All members of the family are characterized by asymmetric data rates (downstream > upstream) and support for the use of an analog telephone (POTS) or ISDN on the same line; this makes the ADSL family an ideal solution for residential subscribers. See the last section of Chapter 5 for a detailed discussion of ADSL.

G.993.1: VDSL. This VDSL flavor exists in two modulation variants: Single Carrier Modulation (SCM or "QAM") and Multi-Carrier Modulation (MCM or "DMT"). It targets both symmetric and asymmetric services on short loops (≤1.5 km). See Chapter 4 for a detailed discussion of VDSL.

The Transport Protocol Specific Transmission Convergence (TPS-TC) layer sits on top of the α(β)-interface and is flavor-agnostic. The TPS-TC transmits and receives protocol datagram units encoded as an octet stream. It contains all the functions necessary to run the specified protocol or family of protocols over the xDSL link, including encoding of events such as start-of-frame and end-of-frame if applicable. The upper termination of the TPS-TC layer is the γ-interface. Different instantiations of the TPS-TC have been defined, the most common of which are ATM-TC, STM-TC, and PTM-TC. The ATM-TC transmits and receives 53-byte cells in ATM; the preferred implementation of the

---

[7] Flavor-Specific Transmission Convergence might have been a more appropriate name.

ATM-TC γ-interface is a Utopia bus. The STM-TC is designed to carry Synchronous Transfer Mode data, typically in multiples of the 64 kbps rate required for traditional telephony lines. The PTM-TC is discussed in more detail in the next section.

Transmission Convergence Sublayer

Packet Transfer Mode

In August 2001, consensus was reached in ITU-T Study Group 15 to add a transmission convergence layer for a "Packet Transfer Mode" to the VDSL Foundation Document (ITU-T Recommendation G.993.1). With this addition, the tradition of having only ATM and STM in xDSL standards ended, and it is no coincidence that this happened in the context of the VDSL project:[8] given the relatively high bit rate supported by VDSL, there is no justification for ATM in terms of Quality of Service or packetization jitter. Moreover, certain prestandard implementations of *Ethernet-over-VDSL* were already available on the market at that time.

The Packet Transfer Transmission Convergence (PTM-TC) is a sublayer bounded by the functional α(β)-interface and the γ-interface, as shown in Figure 3.3. Its main function is to preserve packet boundaries during transmission, in order to provide a packet pipe transparent to the higher layers (collectively referred to as the *packet entity* or *PTM entity*), and unaware of the protocols used. To this end, the PTM-TC uses high level data link control (HDLC)[9] encapsulation in synchronous (or octet-based) mode. HDLC encapsulation was chosen for its simplicity, and for the fact that it was already part of the VDSL standard as the protocol for the embedded overhead channel (eoc).

The HDLC frame structure is illustrated in Figure 3.4. HDLC encapsulation prepends an opening flag (0x7E), an address octet (all stations: 0xFF), and a control octet (unnumbered data: 0x03) and appends a 16-bit or 32-bit CRC and a closing flag (0x7E). A single flag may be used to delimit two frames that immediately follow each other. In the absence of transmit data, a continuous stream of flags is transmitted. HDLC control signalling between the end stations is not used in the PTM-TC (i.e., both stations may transmit independently of each other).

---

[8] Packet Mode for ADSL had already been specified by the ADSL Forum in 1997 [42], but this was never formally adopted in any ADSL standards.

[9] HDLC is specified in the ISO/IEC 3309 standard.

	MAC Control
Packet Entity	MAC
	Reconciliation
Physical Interface	MII
Adaptation Layer	PCS
	γ
PTM-TC	TC Sublayer
	α(β)
PMS-TC	PMA
	I
PMD	PMD
	U

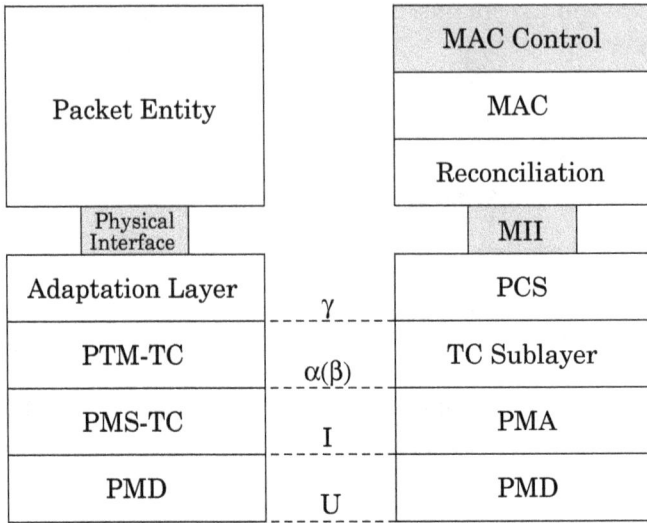

Figure 3.3   Ethernet-over-xDSL reference model (shaded portions are optional).

Flag 0x7E	Address 0xFF	Control 0x03	PAY-	LOAD	CRC-16	Flag 0x7E

Figure 3.4   HDLC encapsulation in the PTM-TC layer.

To avoid false packet ends, *byte stuffing* is used. Any occurrences in the data stream of the 0x7E octet are replaced by 0x7D5E and any occurrences of the escape octet 0x7D are replaced by 0x7D5D. Because byte stuffing replaces a single octet by two octets, the actual overhead generated by HDLC will vary between frames. A worst-case frame, containing only 0x7D and 0x7E octets, will double in size after byte stuffing. On average, however, the overhead introduced by flag insertion for random data is 0.78 percent.

Link transparency is guaranteed by adding a 16-bit cyclic redundancy check (CRC), known as CRC-CCITT, to each HDLC frame. When better error-protection is required, a 32-bit CRC may optionally be used. For a more detailed discussion of CRC calculation, see the section "64/65-Oclet Encapsulation."

The γ-interface defined for the PTM-TC (shown in Table 3.2) is specified in such a way that no frame buffer is required in the PTM-TC sublayer. This is accomplished by placing the PTM-TC in control of the data flow over the γ-interface. Although the higher layers control the clock signals Rx_Clk and Tx_Clk, the PTM-TC sublayer may choose during which clock signals it will transmit or receive an octet and on which clock signals it will be idle by asserting the synchronization signals Rx_Enbl and Tx_Enbl at the appropriate time. This architecture allows the PTM-TC to transmit and received data over the γ-interface at the rate at which this data are supplied or requested by the PMS-TC layer over the $\alpha(\beta)$-interface, leaving room for some extra idle cycles needed for insertion and removal of stuff bytes and CRC.

TABLE 3.1   The PTM-TC γ-Interface[10].

Flow	Signal	Description	Direction
		Transmit Signals	
Data	Tx_PTM	Transmit Data	PTM → PTM-TC
Control	Tx_Enbl	Asserted by the PTM-TC; indicates PTM may push data to the PTM-TC	PTM ← PTM-TC
Control	Tx_Err	Errored transmit packet (request to abort)	PTM → PTM-TC
Sync	Tx_Avbl	Errored transmit packet (request to abort)	PTM → PTM-TC
Sync	Tx_Clk	Clock signal asserted by the PTM entity	PTM → PTM-TC
Sync	Tx_SoP	Start of the transmit Packet	PTM → PTM-TC
Sync	Tx_EoP	End of the transmit Packet	PTM → PTM-TC
		Receive Signals	
Data	Rx_PTM	Receive data	PTM ← PTM-TC
Control	Rx_Enbl	Asserted by the PTM-TC; indicates PTM may pull data from the PTM-TC	PTM ← PTM-TC
Control	Rx_Err	Received error signals including FCS error, Invalid Frame, and OK	PTM ← PTM-TC
Sync	Rx_Clk	Clock signal asserted by the PTM entity	PTM → PTM-TC
Sync	Rx_SoP	Start of the receive Packet	PTM ← PTM-TC
Sync	Rx_EoP	End of the receive Packet	PTM ← PTM-TC

Note that the PTM-TC may only retrieve data from the packet entity when Tx_Avble is asserted. The packet entity must not assert Tx_Avble unless it has a complete packet available for transmission; this avoids getting in the undefined situation, which occurs when

---

[10] Reproduced with the kind permission of ITU.

Tx_Avble is temporarily deasserted during the transmission of a packet. Unlike the MII signals TX_EN and RX_DV (see Figure 2.1), Tx/Rx_Avble are not used to indicate the boundaries of a packet (Tx/Rx_SoP and Tx/Rx_EoP are used for this purpose), but to indicate the availability of data (at least one packet available in transmit direction, at least one octet available in receive direction).

From PTM-TC to EFM-TC

The PTM-TC is a generic layer, independent of protocol-specific parameters such as maximum frame length and frame structure. It was specifically designed to accommodate any frame-based protocol by means of an additional adaptation layer on top of the PTM-TC γ-interface. The PCS of EFM Copper could have been defined in that way (PTM-TC + Ethernet Adaptation Layer), as was originally proposed to the Task Force by the author and others, but in the end, the arguments for doing things differently outweighed the arguments for sticking to the existing ITU-T recommendation.

The PTM-TC takes care of packet delineation (the "flag" bytes, with a value of 0x7E, signal the start or the end of a packet) and link transparency (a 16-bit CRC is used for error detection). The only requirement left for an Ethernet-over-xDSL adaptation layer would be rate matching between the Media Access Control (MAC) and the PHY, and (optional) aggregation of multiple loops. The essential proposals for MAC-PHY rate matching and loop aggregation were presented to the Task Force at the Raleigh meeting in January 2002, and the copper part of the EFM standard could have been finished there and then. Apart from speeding up the standardization process, the advantage of an EFM specification based on the existing PTM-TC and γ-interface, would have been that the adaptation layer could have been applied to any of the xDSL flavors standardized by ITU-T. In other words, one effort would have given the world standards for Ethernet-over-VDSL, Ethernet-over-SHDSL, Ethernet-over-ADSL, and Ethernet-over-any-future-xDSL, as suggested by the architecture in Figure 3.3.

At several times in its history, the 802.3 Working Group has had to choose encapsulation methods for its different components. HDLC had been proposed and rejected in the past, mainly due to the fact that the HDLC frame delimiters are not sufficiently robust against errors: a single bit error can convert a data byte into a false delimiter or vice versa. The 16-bit CRC (or even the optional 32-bit variant) is not sufficiently powerful to exclude the possibility of false packets being for-warded to the MAC.

For these reasons, the EFM Task Force chose to go forward with a new TPC-TC (simply called the "TC sublayer" in the standard, and "EFM-TC" in this book), in addition to a Physical Coding Sublayer (PCS) that serves as an adaptation layer between the EFM-TC γ-interface and the Media Independent Interface (MII).

The unfortunate result is that there are now two *different* stan-dard-compliant ways to transport Ethernet frames over a VDSL link: ITU-T Recommendation G.993.1 in Packet Transfer Mode, and IEEE 802.3ah 10PASS-TS. Both will probably continue to exist in the mar-ket for a while, although it is expected that future Ethernet-over-xDSL specifications will reference the EFM-TC, or a modified form thereof.

The functions of the EFM-TC and the PCS are described in the remainder of this section. An overview of the resulting architecture is shown in Figure 3.5.

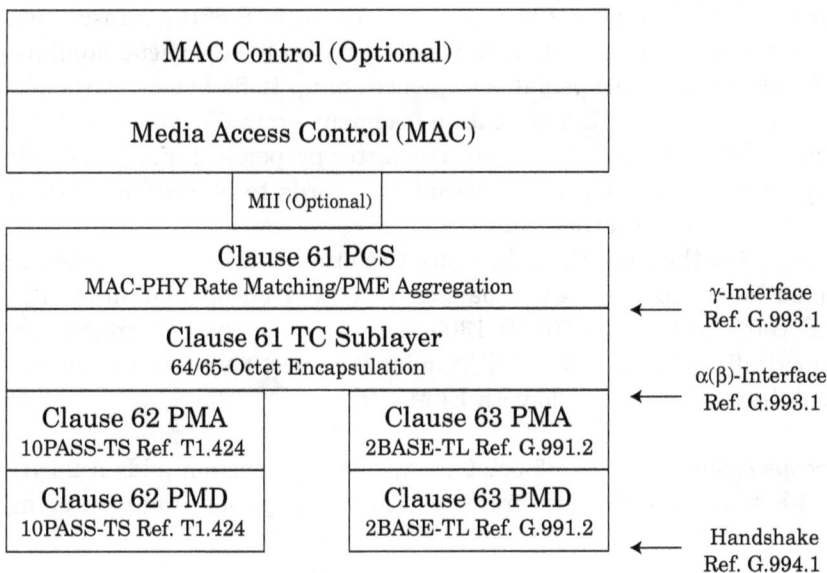

| MAC Control (Optional) |
| Media Access Control (MAC) |
| MII (Optional) |

Clause 61 PCS	
MAC-PHY Rate Matching/PME Aggregation	
Clause 61 TC Sublayer	
64/65-Octet Encapsulation	
Clause 62 PMA	Clause 63 PMA
10PASS-TS Ref. T1.424	2BASE-TL Ref. G.991.2
Clause 62 PMD	Clause 63 PMD
10PASS-TS Ref. T1.424	2BASE-TL Ref. G.991.2

← γ-Interface Ref. G.993.1

← $\alpha(\beta)$-Interface Ref. G.993.1

← Handshake Ref. G.994.1

Figure 3.5 EFM copper reference model.

64/65-Octet Encapsulation

Overview. Generally speaking, frame delineation is accomplished in one of two ways:

- By defining a delimiting symbol, any occurrences of which must be removed from the payload to avoid false frame endings (by bit or byte stuffing, as in HDLC, or by translation into a larger codeword space, as in 8B/10B coding); or
- By indicating the length of the frame in a header, which is added to the payload.

64/65-octet encapsulation is a hybrid of these methods, in the sense that it uses a 65-octet codeword space to encode 64-octet data clusters; the 65-octet codeword can contain an end-of-frame indication.

The 64/65-octet[11] encapsulation function was designed to replace the PTM-TC. It fulfills the requirements of frame delineation and link transparency (error detection), without the problem of variable overhead, which made HDLC unacceptable to the Task Force. In early drafts, a scrambler was provided to improve synchronization with the codeword boundaries,[12] but this scrambler was later abandoned in favor of a synchronization feedback mechanism.

The 64/65-octet encapsulation mechanism is a completely new feature of EFM. It must not be confused with the 64B/66B encapsulation used in Gigabit Ethernet, with which it has only superficial similarities. The former encapsulates sequences of up to 64 bytes—parts of a MAC frame or a PME aggregation fragment—into 65-byte units to be handed off to the $\alpha(\beta)$-interface. The latter prepends 2 bits to a 64-bit sequence to allow space for special codewords to be used on certain optical Gigabit Ethernet links.

To add to the confusion, there are two other bitwise coding schemes called "64B/65B," defined as part of the GFP-T encapsulation in ITU-T Recommendation G.7041/Y.1303 and as part of the PCS in the current draft of the 10GBASE-T standard, respectively. These schemes don't have anything to do with EFM either.

Encapsulation.   The 64/65-octet encapsulation function adds a 2-byte or 4-byte CRC to the data frame or PME aggregation fragment being

---

[11] The word "octet" was chosen by the EFM Task Force to denote 8 bits of data. "Octet" is more accurate than the more familiar "byte," because the word "byte" is sometimes used for symbols of more or less than 8 bits. Moreover, the abbreviation "64B/65B" initially used to denote the EFM Copper encapsulation technique (where "B" stands for "byte") caused confusion because, in existing similar abbreviations "B" is used for "bit" (4B/5B, 8B/10B, 64B/66B). In this book, I will use the word "byte" exclusively for 8-bit data symbols.

[12] For the interested reader, the scrambler was based on polynomial $G(x) = 1 + x^{18} + x^{23}$.

processed,[13] creating a "TC frame," and breaks up this TC frame into 64-byte chunks. The 64-byte chunks are then translated into 65-byte codewords.

The extra byte in 64/65-octet encapsulation is a sync byte, used to indicate whether the remaining bytes consist of data belonging to a single TC frame (value 0x0F), or whether there is a TC frame boundary or a number of idle bytes present (value 0xF0). When the sync byte is 0xF0, it is either followed by an idle byte or by a code byte indicating the position of the last byte of the current TC frame, according to the translation shown in Table 3.3.

TABLE 3.3   TPS-TC Encapsulation Codeword Translation

Codeword	Value
D_n	0x00 – 0xFF
S	0x50
Z	0x00
Y	0xD1
C_n $(n \in [0,63])$	$n + 16 + ((n+1) \bmod 2) \times 128$
Sync	0xF0
Sync All-Data	0x0F
Pre-emptive Sync*	0xF5
Pre-emptive Sync All-Data*	0xAF

Symbols marked with an asterisk are not defined in EFM; they may only be used in systems complying to ITU-T recommendations supporting the EFM-TC.

The bytes following the last byte of the current TC frame, as indicated by the code byte, can consist of any of the following:

- Idle bytes only
- A number of idle bytes (possibly zero), followed by a single start byte and data belonging to the next data frame

The different types of 64/65-octet codewords that can occur are summarized in Table 3.4.

---

[13] The EFM standard uses the word "fragment" to designate either a data frame or a PME aggregation fragment, because the description of the TC is not affected by the presence of the optional PME Aggregation Function.

TABLE 3.4   Examples of 64/65-Octet Encapsulation

Codeword Type	Sync	B_1	B_2	B_3	...	...	...	...	B_{63}	B_{64}
All Data	0x0F	D_0	D_1	D_2	...	...	...	...	D_{62}	D_{63}
End of Frame	0xF0	Ck	D_0	D_1	...	D_{k-1}	Z	...	Z	Z
All Idle	0xF0	Z	Z	Z	...	...	...	...	Z	Z
SoF While Idle	0xF0	Z	...	Z	S	D_0	D_1	...	D_{k-1}	D_k
SoF While Xmit	0xF0	C_k	D_0	D_1	...	D_{k-1}	Z	...	S	D'_0
Idle Out of Sync	0xF0	Y	Z	Z	...	...	...	...	Z	Z

The normal sequence of codewords is illustrated in the left-hand part of Figure 3.6, where transitions are indicated by the value of the sync byte that starts the next codeword. The right-hand part of Figure 3.6 represents an addition specified in the variant of the EFM-TC, as adopted by ITU-T (this addition is *not allowed* in an IEEE Std. 802.3ah compliant implementation). It contains a second instance of the TC state machine (a nearly exact copy of the first one), which is allowed to interrupt the first one at any codeword boundary, thus allowing frames with a higher priority (pre-emptive frames) to be transmitted when transmission of a lower-priority frame has already commenced. The two state machines represent two distinct logical data paths, each with their own γ-interface. The different γ-interfaces connect to distinct higher-layer entities, potentially ending at two distinct MIIs. Practical ways of dealing with such a multiple-path architecture are discussed in the section "Latency and Pre-emption."

Barring errors, the state of the TC sublayer at the receiver side can be characterized as either in-frame or out-of-frame. The TC transitions to the in-frame state when, and only when, an S byte is detected (on reaching synchronization, the transmitting TC must start out with a 0xF0 sync byte and at least one Z byte followed by an S byte). The TC transitions to the out-of-frame state after detecting an End of Frame codeword (sync byte 0xF0) and counting the appropriate number of remaining data bytes, as indicated by C_n.

The CRC added by the 64/65-octet encapsulation function is referred to as "TC-CRC," to distinguish it from the frame check sequence (FCS) already present in the MAC frame. The generating polynomials for the TC-CRC were specifically selected to be different from the generating

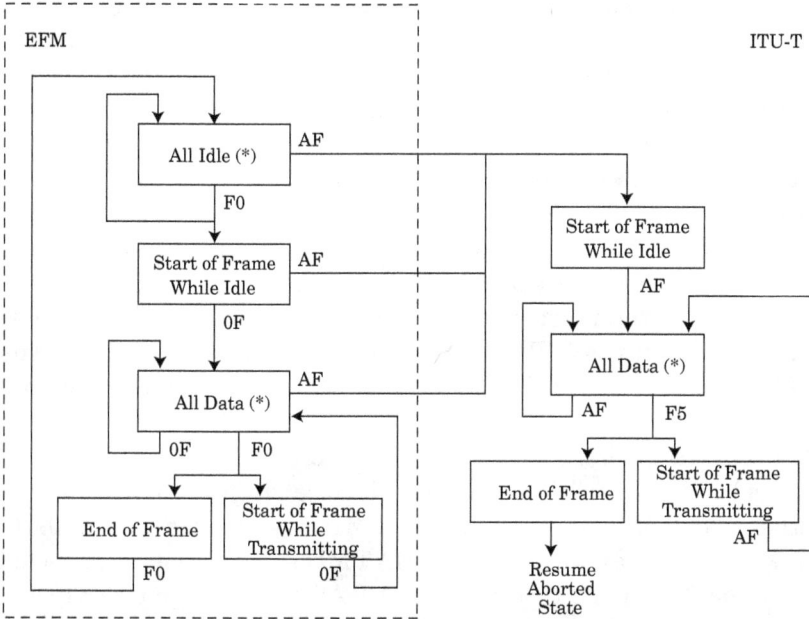

Figure 3.6 Normal sequence of codewords generated by the EFM TC-sublayer, including the optional pre-emptive state machine defined in ITU-T. (Codewords marked with an [*] may be skipped, transmitted once, or transmitted multiple times.)

polynomial of the CRC in the MAC frame, to improve the error detection probability.

When the TC-CRC is used with a 2BASE-TL PMA and PMD, it is defined by the following generating polynomial.[14]

$$x^{32} + x^{28} + x^{27} + x^{26} + x^{25} + x^{23} + x^{22} + x^{20} + x^{19} + x^{18} + x^{14} + x^{13} + x^{11} +$$
$$x^{10} + x^9 + x^8 + x^6 + 1 = (x + 1)(x^{31} + x^{30} + x^{29} + x^{28} + x^{26} + x^{24} + x^{23} +$$
$$x^{22} + x^{18} + x^{13} + x^{10} + x^8 + x^5 + x^4 + x^3 + x^2 + x + 1) \tag{3.5}$$

In case the TC sublayer is used with a 10PASS-TS PMA and PMD, the traditional 16-bit CRC-CCITT is used as TC-CRC, defined by the following generator polynomial.

$$x^{16} + x^{12} + x^5 + 1 \tag{3.6}$$

---

[14] Polynomials reprinted from IEEE Std. 802.3ah-2004, "Ethernet in the First Mile (EFM)," by IEEE. © 2004 IEEE. All rights reserved.

10PASS-TS can do with a shorter, less powerful CRC, because it has a lower undetected error ratio for the same nominal BER, thanks to the presence of Reed-Solomon forward error correction. For this benefit to be noticeable at the TC sublayer, the PMA needs to signal the presence of uncorrectable errors to the TC sublayer. The EFM standard does not specify how this information is to be passed up from PMA to TC, but it can logically be considered as an additional signal on the $\alpha(\beta)$-interface. The capability to detect uncorrectable errors is not common in existing VDSL implementations.

Incidentally, the nominal SNR margin used in 2BASE-TL is also 1 dB lower than the SNR margin used in 10PASS-TS (5 dB versus 6 dB), which gives 10PASS-TS another BER benefit under nominal conditions.

If the EFM-TC is implemented as a stand-alone component or IP block (a VHDL or Verilog module, for instance), to be combined with an off-the-shelf VDSL or SHDSL PMA/PMD module, the TC needs to "know" in some way whether the PMA/PMD stack that it is connected to is of the 2BASE-TL type or of the 10PASS-TS type, in order to calculate the appropriate TC-CRC. The $\alpha(\beta)$-interface between the PMS-TC and the TPS-TC, as defined in existing VDSL and SHDSL standards, did not offer any way to provide this information. The EFM Task Force therefore added the *PMA_PMD_type* signal (better known as the "flavor byte") to the $\alpha(\beta)$-interface, an 8-bit–wide signal identifying what lies beneath as 10PASS-TS-O, 10PASS-TS-R, 2BASE-TL-O, or 2BASE-TL-R. Of the 126 remaining flavors, one block of 4 was reserved for allocation by ATIS T1E1.4, to allow them to reuse the EFM-TC with other standardized PMA/PMD combinations. Further block reservations may happen in the future. When the EFM-TC is combined with ADSL in this way, the 16-bit TC-CRC is used.

The mathematically most straightforward (but not the most efficient) way to implement the CRC function is by means of a shift register, preinitialized in the all-1 state, into which the data bytes are shifted at the x^0 side, LSB first. Whenever the bit shifting out of register at the x^{32} side (2BASE-TL) or x^{16} side (10PASS-TS), the carry bit is equal to 1, and the generator polynomial (represented as a string of bits) is added[15] to the contents of the shift register. The bytes of the TC-CRC are complemented and appended to the scrambled data stream. At the receiver side, an identical implementation will lead to

---

[15] Addition is defined as a bitwise exclusive-or operation.

final shift register contents equal to 0x1C2D19ED (2BASE-TL) or 0x1D0F (10PASS-TS) when the TC frame was received without errors.

Synchronization.   The receiving side of the TC sublayer achieves synchronization by detecting 4 consecutive sync bytes (0xF0 or 0x0F) with no alternative sequence of more than 2 sync bytes in the same period. If everything goes right from the start, synchronization can be achieved after the first $3 \times 65 + 1 = 196$ bytes.

When a sync byte is expected but not received, the TC sublayer starts *freewheeling* (but it is still considered synchronized). During the freewheeling state, a single expected sync byte will bring the TC sublayer back to the normal synchronized state. The goal of the freewheeling state is to provide a buffer against burst errors; the bursty nature of noise on telephone wires will cause errors to occur in concentrated bursts, wiping out a number of consecutive symbols, without necessarily losing byte synchronicity. As long as byte synchronicity is preserved, the receive state machine will know where to look for sync bytes. It will thus be able to temporarily maintain an "error-aware" synchronized state (freewheeling), during which data transmission may continue. However, if a wrong byte is received instead of an expected sync byte, this must be signalled to the PCS by means of the *Rx_Err* signal on the γ-interface. If the receive state machine was out-of-frame, the assertion of the *Rx_Err* signal will not affect the validity of any data received before the detection of the coding violation. If the state machine was in-frame, the *Rx_Err* signal will invalidate the current frame or fragment.

If 3 more expected sync bytes are missed in the freewheeling state, the TC will lose synchronization and stop transferring data to the PCS, while maintaining knowledge of the expected sync byte position. In this "freewheeling out-of-sync" state, a single expected sync byte brings the TC sublayer back into the normal synchronized state. If 4 more expected sync bytes are missed in the "freewheeling out-of-sync" state, the TC sublayer finally erases all knowledge of codeword boundaries and starts looking for 4 consecutive sync bytes again. Thus, between the first missed sync and the total loss of synchronization leading to the "looking" state, $7 \times 65 = 455$ bytes have to be received.

Loss of synchronization, either local or remote, is signaled to the PCS over the γ-interface to keep it from sending down data while the TC sublayer is out-of-sync. For this purpose, the *TC_link_state* signal was added to the γ-interface (from the TC sublayer to the PCS). Figure 3.7 illustrates the relation between the different relevant state

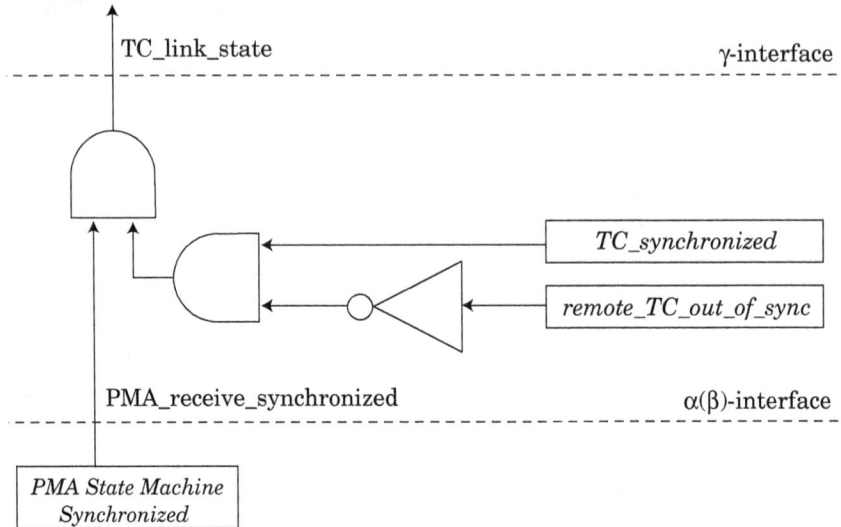

Figure 3.7   Relation between state machine variables and interface signals concerning synchronization.

machine variables (shown in italics inside boxes) and interface signals (shown next to arrows). The pseudo-variable *PMA State Machine Synchronized*, which is logically ANDed into the *TC_link_state* signal, represents all states in which the PMA is capable of transmitting and receiving data, regardless of the actual flavor-specific details. For 10PASS-TS, this corresponds to the internal VDSL link state and timing diagram being in state *STEADY_STATE_TRANSMISSION*; for 2BASE-TL, this signal corresponds to the inverse of the LOSW bit, which indicates a Loss of Sync Defect.

When the TC sublayer is not synchronized (*TC_synchronized* is false), it must not send any data; it will send "all idle out-of-sync" codewords consisting of a *Y* followed by regular *Z* idle bytes to indicate its loss of synchronization. In response, the remote TC sublayer shall assert *remote_TC_out_of_sync* and send regular idle codewords, which should facilitate the local TC sublayer's transition back into synchronization.

Difficulties may arise when transmission of a frame is already in progress when synchronization is lost. To avoid ambiguities, the state diagrams provide very specific rules for these situations.

The transmit state diagram clarifies that:

- Transmission of a frame from PCS to TC must never start when *TC_link_state* is false; and

- If transmission of a frame or fragment is in progress when *TC_link_state* becomes false, no more octets will be pulled from the PCS to the TC, because the TC moves to a state where it sends "all idle" or "all idle out-of-sync" codewords, after filling up the codeword in progress with Z bytes. The PCS is responsible for discarding whatever data of the frame or fragment in progress was left in its transmit buffer.

Similarly, the receive state diagram clarifies that:

- Receipt of a frame or fragment will never start when *TC_link_state* is false, as no valid data are coming in; and
- If receipt of a frame or fragment is in progress when *TC_link_state* becomes false (which implies that *TC_synchronized* has become false), the flow of valid incoming data is broken, and the TC will signal the premature ending of the frame of fragment in progress to the PCS by means of the *Rx_Err* signal on the γ-interface.

Both *TC_synchronized* and *remote_TC_out_of_sync* are accessible through management, as explained in the last section of this chapter.

A 2BASE-TL symbol contains up to 4 bits of data, while a single 10PASS-TS Discrete Multi-Tone (DMT) symbol can contain thousands of bytes of data (depending on the configured bitrate). As a result, a 10PASS-TS link can lose and regain synchronization within the space of a single DMT symbol, or lose synchronization in one symbol and regain it in the next. In these cases, the transmission of "all idle out-of-sync" codewords to the peer will cause the peer to transmit idle codewords *at least two DMT symbols later* (the data inside a DMT symbol only become available to the PMA and TC sublayers when the entire symbol has been received and demodulated; the FEC block and interleaver will cause additional delay). Not only are these idle codewords quite useless—the receiver has already acquired synchronization—they also cause perfectly good data to be flushed from the peer's transmit buffers when it transitions to the idle state. Therefore, it is recommended (though this is not stated in the standard) to extend the duration of the freewheeling state to cover at least two entire 10PASS-TS symbols at high bitrates.

Error handling.   If an incorrect value of C_n is received, the codeword shall be ignored, and the TC skips to the next codeword. This will certainly kill the TC frame that was being received, because the ending of this frame will be discarded. It may also kill the next TC frame, which might have started within the same codeword. An adventurous implementation might manage to save both TC frames by carefully analyzing the contents of the codeword; it should look like data, followed by zero or more Z bytes, possibly followed by an S byte and more data. If this is the case, it seems straightforward to detect the start of the second TC frame; however, valid data might coincidentally look exactly like such a pattern. The TC-CRC register can help detect the end of the first data frame, as it will display a known value when it has processed a data frame with a correct CRC sequence appended to it.

Considering that all speculation about the TC frame boundary is risky at best in the case of an incorrect received C_n value, a more foolproof error recovery scheme is provided in the standard. After an incorrect received C_n value or an incorrect received sync byte (assuming that this "missed sync" does not cause synchronization to be lost), the only way to be certain that no false frame boundaries are being introduced by the TC sublayer is to wait for a valid 0xF0 sync byte, followed by a valid C_n. After the data frame ends, an S byte marks the start of the next valid data frame.

All coding errors as well as CRC errors are monitored and counted by management objects, which are accessible, as explained later in this chapter.

Closing remarks.   As an additional point of interest on the TPS-TC, let us explore the question of *where the TPS-TC ends*. From an architectural point of view, and using the terminology of the ITU-T Recommendations, the TPS-TC ends where the transport protocol takes over. In the case of Ethernet, this is at the MII. The whole of the PCS and the TC sublayer defined in Clause 61 of the EFM standard could be seen as the specification of a new TPS-TC: the Ethernet-TC. The MII would then become an implementation of the Ethernet-TC–specific γ-interface. From a legalistic point of view, the TPS-TC ends at whichever place the SDO draws a line and labels it γ-interface. In this view (which is the only standard-compliant view), the EFM TPS-TC contains only the encapsulation function, while the PCS should be considered "adaptation." In the case of the ATM-TC, the transport protocol–specific interface (Utopia) and the γ-interface coincide, which is obviously architecturally more pure. In ATM, all necessary adaptation is done above the γ-interface, by means of "ATM Adaptation Layers" (such as AAL5).

A Note on Bit Ordering

When octet-oriented data are transmitted serially, a certain convention must be adopted as to which bit of the octet is transmitted first. Unfortunately, different practices exist in different fields of data transmission. As a result, a standard that resides on the border of different fields, such as EFM Copper—a marriage of Ethernet and xDSL—is bound to have some ambiguities about the order of bit transmission. To make things worse, different encoding algorithms, such as scramblers and CRC generators, may operate on serial data in different ways. The goal of this note is to clarify bit ordering issues in EFM Copper.

Media-independent interface.   The MII transmits and receives octet-data in nibbles (groups of four bits). The first nibble contains the four least significant bits; the second nibble contains the four most significant bits. The least significant bit of the first nibble will be denoted LSB or left-most bit; the most significant bit of the second nibble will be denoted MSB or right-most bit.

No serial processing is performed on the data between the MII and the γ-interface; the only functions residing here are MAC-PHY Rate Matching and PME Aggregation.

γ-interface.   The γ-interface may transmit and receive octet-data serially, in which case the LSB as defined above is transmitted/received first.

The sublayer between the γ-interface and the $\alpha(\beta)$-interface is the TPS-TC sublayer (or TC sublayer, for brevity). In this layer, a TC-CRC is calculated, and encapsulation overhead is added. If implemented as a serial processor, the CRC operates on the LSB first, by shifting it into the x^0 side of the shift register.

The result of the CRC calculation can be found in the CRC register when a complete data fragment has been processed. This result must be read backwards, such that:

- For 10PASS-TS, the x^{15} bit is considered as the LSB of the first TC-CRC octet and the x^0 bit is the MSB of the second TC-CRC octet; and
- For 2BASE-TL, the x^{31} bit is considered as the LSB of the first TC-CRC octet and the x^0 bit is the MSB of the fourth TC-CRC octet.

$\alpha(\beta)$-interface.   The $\alpha(\beta)$-interface may transmit and receive octet-data serially, in which case the LSB as defined above is transmitted/ received first. The conventional way to specify the $\alpha(\beta)$-interface, is to have it transmit the MSB first. This traditional convention is not adhered to in EFM Copper! (We can make it look like the convention is adhered to, however, by renaming the LSB as defined above to MSB on the $\alpha(\beta)$-interface. This does not change the fact that the LSB as defined at the MII is processed first in any serial processing below the $\alpha(\beta)$-interface, such as the PMA scrambler and the PMA CRC calculation.)

I-interface and below.   The I-interface is a purely abstract octet interface. The issue of bit ordering doesn't arise on the I-interface. All functions specified below the I-interface use the designation of LSB and MSB, which shall be interpreted to be consistent with the definition of LSB and MSB on the MII. (This is also consistent with the pro-forma renaming of LSB to MSB on the $\alpha(\beta)$-interface, and then executing the bit reordering instructions specified in T1.424 Section 9.3.)

The EFM-TC in Other Standards

In the same way in which EFM Copper relies on other standards for the specification of the PMD and PMA of its transceivers, other standards have come to rely on EFM Copper for the specification of certain TPS-TC- and PCS-related functions. The advantage of using common sublayer interfaces in the different xDSL standards is that individual sublayers can easily be exchanged between standards.

From the onset, it has been an explicit objective of the EFM Task Force to specify a TC layer with a γ-interface that is functionally compatible with the one specified for the PTM-TC in ITU-T Recommendation G.993.1. As the existing PTM-TC itself was replaced in EFM by 64/65-octet encapsulation, functional compatibility with the $\alpha(\beta)$-interface became an additional goal (though this was never formally stated). This means that the EFM-TC can be used in any system with a byte-synchronous interface that is capable of pushing data to and pulling data from the EFM-TC.

ITU-T Study Group 15 has agreed to add a "PTM-TC based on 64/65-octet encapsulation" to its recommendation G.992.3 (ADSL2), with the intent of referencing this sublayer in subsequent editions of other xDSL standards (VDSL2, [E]SHDSL). Unfortunately, this new PTM-TC is not identical to the EFM-TC.

As the ITU-T packet-mode TPS-TC needs to be sufficiently generic to support any packet protocol—not just Ethernet—the minimum packet size of 64 octets imposed by 64/65-octet encapsulation is a serious obstacle. (For example, if native IP is transported over an xDSL link, the encapsulation protocol should be able to deal with a different range of frame sizes. Such an architecture may be used for a routed xDSL network, in which both the modem and the access node have IP awareness.) Shorter frames are supported in the "PTM-TC based on 64/65-octet encapsulation" by allowing a frame to start with a C_kS sequence[16] any time during the idle state.

Another ITU-T modification is to allow codewords belonging to high-priority TC frames to get in between the codewords of a low-priority TC frame that is already being transmitted (preemption). A similar suspend/resume mechanism already exists in PPP.[17] This is achieved in the PTM-TC by defining two new sync bytes (coded as shown in Table 3.3 and used as shown in Figure 3.6). The combination of PME aggregation and 64/65-octet encapsulation with suspend/resume leads to a sublayer that could be characterized as "ATM light," where the streams of 65-byte codewords are streams of "cells" that can be demultiplexed and interleaved in the way ATM connections can.

Although the enhancements described here have been specified in such a way that an enhanced receiver will correctly interoperate with a transmitter that has a standard EFM-TC, the ITU-T's "PTM-TC based on 64/65-octet encapsulation" must be regarded as a new TPS-TC (i.e., a new set of compliance rules, different from those in Clause 61 of IEEE Std. 802.3ah). ITU-T's "PTM-TC based on 64/65-octet encapsulation" seems to come in addition to the existing HDLC-based PTM-TC, and not in its place. Hence, there will be one more way to transport Ethernet over xDSL, instead of one less....

Committee T1E1.4 references the PME Aggregation function in the EFM standard for its Ethernet bonding standard, but as no packet transport layer had been specified in any of the T1E1.4 xDSL standards, they decided to adopt the EFM-TC sublayer along with it. Unlike ITU-T Study Group 15, T1E1.4 adopted the EFM-TC without modifications, including the $\alpha(\beta)$-interface with the flavor byte. A block of transceiver-type codepoints (possible values for the flavor byte) have been reserved in the EFM standard for use by T1E1.4.

---

[16] The sequence $Z\ C_k$ is still being considered as an alternative "start-of-short-frame" code.

[17] See RFC 1990 for details on the Multi-Link protocol; see RFC 2686 and RFC 2687 for details on Real-Time extensions for PPP.

Physical Coding Sublayer

Aggregation Basics

Several methods exist to aggregate multiple physical transceivers into one logical transceiver with a higher bitrate. The layered architecture of xDSL and Ethernet transceivers, with clearly identified interfaces between the sublayers, is very aggregation-friendly; to allow aggregation, it suffices to either add the aggregation function to an existing sublayer (which will then interface with several lower-layer instances instead of just one), or to specify a new aggregation sublayer, which presents the same interface to the higher layer as the aggregated lower layer instances present to itself.

For Ethernet, there is a mechanism called *link aggregation* (also known as 802.3ad, specified in Clause 43 of IEEE Std. 802.3), which combines a number of identical full-duplex PHYs of a system into a single "fat" Ethernet pipe. This is done by dividing the stream of transmit frames from the attached MAC Client or MAC Relay into conversations, which are sets of frames sharing the same Source Address and Destination Address. By assigning all the frames of a given conversation to the same link, frame reordering and duplication within a conversation can be avoided.[18] The Aggregator is governed by a Link Aggregation Control function, which may be configured manually through dedicated management variables or automatically by means of the Link Aggregation Control Protocol (LACP). LACP continually monitors the eligibility of links to participate in the aggregation; as such, it provides a simple and rapid redundancy mechanism for links between multi-port Ethernet elements.

Where Ethernet PHYs typically come in bitrates that are powers of 10, link aggregation provides a way to create links with bitrates that are integer multiples of the original rates. In ATM architectures, Inverse Multiplexing for ATM (IMA) provides a similar functionality. In both cases, an aggregation layer is placed between the MAC/ATM service interface of the transceivers and the MAC/ATM client. The client is unaware that its data are being processed by multiple transceivers, and the transceivers are unaware that the data they are processing is only part of a larger whole.... The aggregation layer hides these details from the entities it interfaces with.

---

[18] Reordering the frames for a given combination of source address and destination address and frame duplication are prohibited by the specification of the MAC Service (see ISO/IEC 15802-1 and IEEE Std. 802.1D); some exceptions are granted for the operation of RSTP.

At lower layers, aggregation is often referred to as "bonding." In its "G.bond" project, ITU-T Study Group 15 has been looking into generic xDSL bonding at lower-layer interfaces, such as the γ-interface (PME Aggregation as defined in EFM Copper is an example of that) or the α(β)-interface. Certain PHYs have a PMD that can operate over multiple pairs (e.g., SHDSL), which can be viewed as a limiting case of aggregation: aggregation at the U-interface (physical medium interface).

PME Aggregation for EFM Copper

The EFM standard specifies a new aggregation function, PME Aggregation, as part of the PCS sublayer. It allows data frames from a single MAC and MII to be fragmented[19] and transmitted over several different PMD/PMA instances. This feature was added in part to satisfy the Ethernet community's desire to be able to offer bitrates in powers of 10. Without aggregation, it would be impossible to offer 10 Mbps with the long-reach PHY, and very hard to offer 100 Mbps with the short-reach PHY. PME Aggregation is an *optional* part of the PCS—in fact, it is the only optional feature—and may be used for both 2BASE-TL and 10PASS-TS.

PME Aggregation is specified for up to 32 links.[20] In the transmit direction, frames are divided into fragments, which receive a Fragmentation Header (see Figure 3.8) and are sent to the TC sublayer. Fragment sizes must be between 64 and 512 bytes (not including the Fragmentation Header), and be multiples of 64, except in the last fragment of a frame. The TC sublayer specified in the EFM standard is actually unable to deal with fragments that are less than 64 bytes long.[21] Hence, *all* fragments must be at least 64 bytes long, even the last one; the fragmentation algorithm must take this into account when deciding the length of the other fragments.

Apart from these restrictions, the actual fragmentation algorithm is up to the implementer.

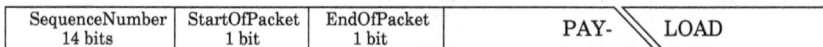

SequenceNumber 14 bits	StartOfPacket 1 bit	EndOfPacket 1 bit	PAY-	LOAD

Figure 3.8 Fragmentation header format.

---

[19] This is thus a major difference with link aggregation.
[20] It is allowed to "aggregate" a single link. In practice, this should only be done if other links are expected to join the aggregate afterwards.
[21] This limitation is remedied by the ITU-T version of the EFM-TC.

At the receiver, the *SequenceNumber*, *StartOfPacket*, and *EndOfPacket* fields in the Fragmentation Header are used to reassemble the frame. The main responsibility of the receiver is to correctly deal with errors, which may occur in fragment reception, sequencing, or reassembly. Depending on whether the errored received frame can be (partially) reassembled, it is either sent to the TC layer with Rx_Err asserted on the γ-interface, or a "garbage frame" is sent instead. The garbage frame consists of 64 bytes of 0x00 (which obviously includes an intentionally faulty FCS).

Given the latency and bitrate variations between the aggregated links, a certain amount of buffering is required at the receiver side. To bound the receive buffer size, two constraints are imposed on the transmitter: The differential latency between any two aggregated links shall not be more than 15,000 bit times, and the highest bitrate ratio between any two aggregated links shall not be more than 4. These restrictions allow the use of a 14-bit sequence number space.

Differential latency between two PMEs is defined as the number of bits, N, that can be sent across the fast link in the time that it takes one *maxFragmentSize* (512 bytes) fragment to be sent across the slow link. The time t_s to transmit a 512-byte fragment across the slow link, expressed in seconds, can be computed as follows, where δ_s is the latency of the slow link in byte times, and Rs is the data rate of the slow link in bytes per second.

$$t_s = \frac{512 + \delta_s}{R_s} \tag{3.7}$$

The latency of the link consists of the propagation delay $\delta_{s,p}$ of the signal over the wire (approximately 5µs/km), the processing time $\delta_{s,PMD}$ in the PMD at each end of the link, the interleaver latency $\delta_{s,int}$ (10PASS-TS only, see below), and the 64/65-octet encapsulation delay $\delta_{s,enc}$.

The interleaver latency in 10PASS-TS is a function of parameters M and I, which are both programmable (see Chapter 4).

$$\delta_{s,int} = M_s \times I_s \times (I_s - 1) \tag{3.8}$$

The differential delay N can then be computed as the number of bits transmitted over the fast link in time t_s, taking into account the latency δ_f (in byte times) and the data rate R_f (in bytes per second) of the fast link.

$$N = 8 \times R_f \times t_s - 8 \times \delta_f$$
$$= 8 \times \left[\frac{R_f}{R_s} \times (512 + \delta_s) - \delta_f \right] \qquad (3.9)$$
$$\leq 15{,}000$$

In 2BASE-TL, the only programmable parameter that affects the latency is the data rate of a link. Assuming an encapsulation delay of 70 byte times, and assuming that the other contributions to the latencies of the slowest and the fastest link are assumed to be much smaller than 512+70 byte times, the only way to satisfy the constraint in Equation 3.9 is by limiting the speed ratio $K = R_f/R_s$ to approximately 3.34.

In cases where the interleaver delay is the dominant term in the latency of both links, the constraint on N can be transformed into a relationship between the slow link's interleaver parameters (M_s, I_s) and the fast link's interleaver parameters (M_f, I_f) for every value of K between 1 and 4.

$$M_s \times I_s \times (I_s - 1) \leq K^{-1} \times [1875 + M_f \times I_f \times (I_f - 1)] - 512 \qquad (3.10)$$

The values of the interleaver parameters are limited to a fixed set.

$$I = 36 \;\; \text{and} \;\; M \in \{2, ..., 52\}$$
$$I = 30 \;\; \text{and} \;\; M \in \{2, ..., 62\} \qquad (3.11)$$

Of the four different combinations of I_s and I_f, we work out the case of $I_s = I_f = 36$, which brings the granularity of the interleaver depth to $36 \times (36 - 1) = 1260$.

$$M_s \times 1260 \leq K^{-1} \times (1875 + M_f \times 1260) - 512$$
$$\Leftrightarrow M_s \leq K^{-1} \times (1.48810 + M_f) - 0.40635 \qquad (3.12)$$

In the limiting case where $K = 1$ and enforcing integer solutions, this leads to the following constraint, which can be met for all legal values of M_f.

$$M_s \leq 1.48810 + M_f - 0.40635$$
$$\Leftarrow M_s \leq 1 + M_f \qquad (3.13)$$

In the other limiting case where $K = 4$ and enforcing integer solutions, this leads to the following constraint, which can be met for $12 \leq M_f \leq 52$.

$$M_s \leq 0.37202 + \frac{M_f}{4} - 0.40635$$

$$\Leftarrow M_s \leq \frac{M_f}{4} - 1 \qquad (3.14)$$

The constraints resulting from the other legal combinations of I_s and I_f are left as an exercise for the reader.

Other Functions of the PCS

As the PCS sits between the TC sublayer (running at wire speed) and the MAC sublayer (running at 100 Mbps), it is formally the place where MAC-PHY rate matching takes place. MAC-PHY rate matching is achieved through carrier deference, as described in Chapter 2.

Another function of the PCS is the removal of the preamble and SFD fields from the MAC frames prior to their fragmentation and transmission to the TC sublayer, and the restitution of the preamble and SFD fields to received MAC frames after their reassembly.

The EFM PCS in Other Standards

Before the completion of the EFM standard, ITU-T Study Group 15 Question 4 adopted an "Ethernet bonding" proposal which consists of a reference to the PME Aggregation function in the EFM standard—then, still a draft. The Ethernet bonding layer runs over the ITU-T's existing HDLC-based PTM-TC or over an adapted EFM-TC.

To support PME discovery between ITU-T xDSL transceivers, ITU-T added the relevant handshaking codepoints to the common part of the G.994.1 codepoint tree, where they are accessible to all G.994.1-compliant transceivers (including 10PASS-TS and 2BASE-TL). As a result, these codepoints could then be removed from the 10PASS-TS- and 2BASE-TL-specific codepoint trees, where they had been defined in earlier drafts.

The new American National Standard for Ethernet bonding, under development in ATIS NIPP-NAI,[22] will use the PME Aggregation function from the EFM standard, running over the EFM-TC.

---

[22] The Network Access Interfaces subcommittee of the Network Interface, Power, and Protection Committee, formerly Committee T1E1.4.

Initialization

Overview

A substantial amount of information has to be exchanged between the CO-modem and a CPE-modem before an EFM Copper link can be brought into a mode where it can transport user data. This is accomplished by means of different initialization mechanisms. The actual sequence of events is different for 2BASE-TL and 10PASS-TS, but both rely on "handshaking" as described in ITU-T Recommendation G.994.1 to start initialization.

ITU-T Recommendation G.994.1 specifies handshaking procedures for xDSL transceivers. These procedures allow for an exchange of essential initialization information between an xTU-O and an xTU-R transceiver, prior to the establishment of an end-to-end data link. The EFM PHYs for point-to-point copper are based on xDSL, and therefore use the same handshaking procedures—with different codepoints.

Handshake messages are exchanged by modulating certain sets of carriers, belonging either to the 4 kHz carrier family (used by SHDSL and QAM-VDSL) or the 4.3125 kHz carrier family (used by all DMT-based xDSL systems). For each of the different xDSL transceivers standardized by ITU-T, there is a mandatory minimum carrier set to be used for handshaking, but use of other additional carrier sets is encouraged. Using additional carrier sets allows detection of other xDSL flavors on the link, which can yield useful troubleshooting information, even when the peer in the handshaking transaction is not an interoperable xDSL flavor. The carrier sets used for EFM PHYs are shown in Table 3.5.

TABLE 3.5   Mandatory Carrier Sets for EFM Copper Handshake

Port Type	Set	Downstream Carriers	Upstream Carriers
2BASE-TL	A4	5 ($\times$ 4 kHz)	3 ($\times$ 4 kHz)
10PASS-TS	V43	257, 383, 511 ($\times$ 4.3125 kHz)	944, 972, 999 ($\times$ 4.3125 kHz)

The symbol rate for G.hs modulation is 4312.5/8 = 539.0625 baud for the 4.3125 kHz family, and 4000/5 = 800 baud for the 4 kHz family. The tones are all modulated simultaneously using differential phase-shift keying (DPSK) modulation (a 180° phase reversal indicates binary 1; a 0° phase difference indicates binary 0).

The G.994.1 detection sequence consists of the phases shown in Figure 3.9.

Initiated by HSTU-R

HSTU-R   HSTU-C

R-SILENT0
          R-TONES-REQ   C-SILENT1

          C-TONES
R-SILENT1

          R-TONE1

          C-GALF1

          R-FLAG1

          C-FLAG1

Initiated by HSTU-C

HSTU-R   HSTU-C

R-SILENT0   C-SILENT1

          C-TONES

          R-TONE1

          C-GALF1

          R-FLAG1

          C-FLAG1

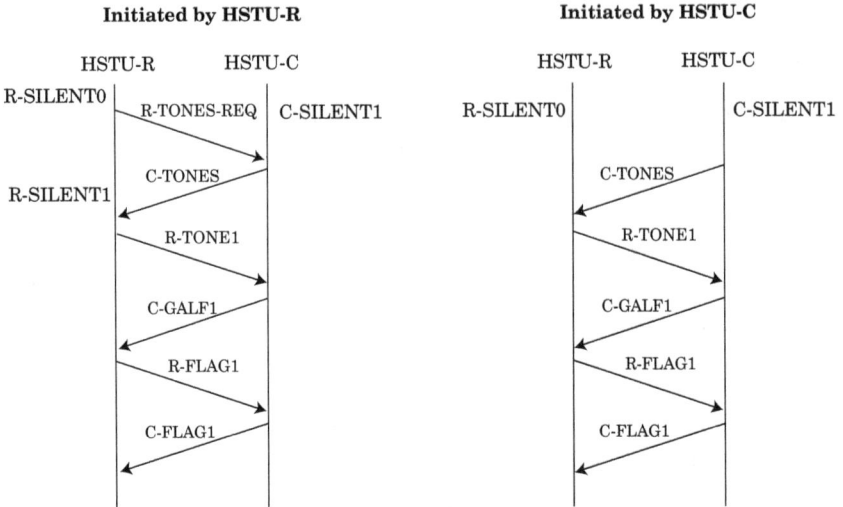

Figure 3.9   G.994.1 detection sequence (duplex).

Once the detection sequence is completed, handshake frames can be exchanged. There are 15 different types of frames defined.

CLR—Capabilities list + request.   Message sent by a CPE-side modem to convey a list of possible modes of operation and to request a similar list (CL message) from the CO-side modem.

CL—Capabilities list.   Message sent by a CO-side modem in response to the reception of a (complete or partial) CLR message, to convey a list of possible modes of operation.

MR—Mode request.   Message sent by a CPE-side modem to request an MS message from the CO-side modem.

MS—Mode select.   Message sent by either modem on the link to request a particular mode of operation.

MP—Mode proposal.   Message sent by a CPE-side modem to propose a particular mode of operation and to request an MS message from the CO-side modem.

REQ-MS—Request MS message.   Message sent by a CO-side modem in response to an MR message to request transmission of an MS message by the CPE-side modem (CO doesn't want to select a mode).

REQ-MR—Request MR message. Message sent by a CO-side modem in response to a (complete or partial) MS message to request the transmission of an MR message by the CPE-modem (CO wants to select the mode).

REQ-CLR—Request CLR message. Message sent by a CO-side modem in response to an MR message, a (complete or partial) MP or MS message to request a CLR message from the CPE-modem (CO wants to exchange capabilities).

ACK(1)—Acknowledge, type 1. Message sent to acknowledge receipt of a (complete or partial) CL message and to end the handshake transaction; or to acknowledge receipt of a (complete or partial) MS message and to initiate the cleardown procedure.

ACK(2)—Acknowledge, type 2. Message sent to acknowledge receipt of a partial CL, CLR, MP, or MS message and to request the transmission of the next frame of the message.

NAK-EF—Negative acknowledge, errored frame. Message sent by either modem on the link in response to an errored frame. It aborts the handshaking session.

NAK-NR—Negative acknowledge, not ready. Message sent by either modem on the link to acknowledge receipt of a (complete or partial) MS message. It ends the handshaking transaction, but indicates that the receiving station wishes to continue the session.

NAK-NS—Negative acknowledge, not supported. Message sent by either modem on the link to acknowledge receipt of a (complete or partial) MP or MS message. It ends the handshaking transaction.

NAK-CD—Negative acknowledge, clear down. Message sent in response to any frame, to indicate that the received information was not understood. It initiates the cleardown procedure.

REQ-RTX—Retransmission message. Message sent to request retransmission of an errored frame.

These messages are used to conduct a number of transactions:

A (MS→ACK(1)). The mode of operation is selected by the CPE modem. The CO modem may refuse the selected mode if there is no common mode of operation possible. In this case, it proceeds to cleardown.

B (MR→MS→ACK(1)). The mode of operation is selected by the CO modem.

C (CLR→CL→ACK(1)). Negotiation of capabilities. To be followed by A, B, or D.

D (MP→MS→ACK(1)). The mode of operation is proposed by the CPE modem. The CO modem is requested to select a mode of operation. The CO modem may refuse to select a mode if no common mode of operation is possible. In this case, it proceeds to cleardown.

In addition to these transactions, a number of "extended transactions" are possible; these occur when the CO-modem doesn't approve of the transaction type initiated by the CPE-modem and overrides this by requesting a specific other transaction by sending REQ-MR, REQ-MS, or REQ-CLR. In this way, the CO-modem is really always the master of the handshaking session.

G.994.1 Frame Structure

The messages CL, CLR, MS, MR, and MP all convey information in three different kinds of fields, known as Identification Fields, Standard Information Fields, and Non-Standard Information Fields. The parameters in these fields are organized as a tree of codepoints. An NPar(n) codepoint represents a certain capability or a requested mode of the transceiver for which no further information is required. An SPar(n) codepoint indicates a capability for which further options or data will be specified in subsequent NPar(n+1) or SPar(n+1) messages.

At the highest level (NPar(1) and SPar(1)) the major capabilities are exchanged: A different SPar(1) bit exists for each of the ITU-T standardized xDSL flavors, as well as for certain transceivers specified by ETSI, Committee T1, and IEEE 802.3 (specifically, 10PASS-TS and 2BASE-TL). At the next level, NPar(2) and SPar(2) bits are used to exchange capabilities specific to the selected port type. Level 3 is the bottom of the tree; only NPar(3) bits are used, which specify additional information for the SPar(2) bits under which they reside.

Identification Field

The G.994.1 Recommendation originally used the Identification Field to convey information about net data rates, data flow characteristics (latencies), and local splitter information. These bits are of an inform-ative nature, and their use is not specified for EFM Copper. A second octet is used to transmit relative power levels for the different carrier sets to be used during the G.994.1 handshake session—it is also used in this fashion in EFM Copper.

Codepoints for PME Aggregation Discovery were originally defined in the EFM Draft. After the PME Aggregation mechanism was adopted by T1E1.4 and ITU-T Q4/15, these codepoints were added to level 1 of the G.994.1 handshake tree (Identification Field), where they are available to all G.994.1 compliant transceivers.

- SPar(1) Bonding
  - NPar(2) Ethernet bonding (used by EFM Copper)
  - NPar(2) TDIM bonding (not used by EFM Copper)
  - SPar(2) PME Aggregation Discovery
    • NPar(3) Clear if same
    • NPar(3) Remote discovery register (48 bits)
  - SPar(2) PME Aggregation
    • NPar(3) PME_Aggregate_register (32 bits)

Standard Information Field

The Standard Information Field for 10PASS-TS and 2BASE-TL are shown in Chapters 4 and 5, respectively. An additional codepoint for a variable silence period was originally defined in the EFM draft. This codepoint was later moved into level 1 of the G.994.1 handshake tree (Standard Information Field), where it is available to all G.994.1–com-pliant transceivers.

- SPar(1) Variable Silence Period
  - NPar(2) Variable Silence Period Length (6 bits used to code length of 10–640 seconds)

Non-Standard Information Field

The Non-Standard Information Field is used "to convey information beyond that defined in this Recommendation" (ITU-T Recommendation G.994.1.) The EFM standard does not specify any use for this field.

Remote Discovery of PME Aggregation

In EFM, the G.994.1 handshake is additionally used to set up an aggregated link composed of several physical EFM links (PME aggregation). This requires several exchanges of information (marked as DISCOVERY and AGGREGATION in the state machine in Figure 3.10), which are initiated by the CO-side. Although either link partner can be the first one to send handshake tones, the actual G.994.1 *transactions* are specified to be initiated by the HSTU-R (CPE-side). Therefore, an EFM transaction involving a request from the CO and a response from the CPE requires two G.994.1 transactions back-to-back; two CLR/CL transactions are executed, wherein the first CL message carries the CO request, and the second CLR message carries the CPE response.

Two registers govern PME aggregation: *PME_Available_register* and *PME_Aggregate_register*. The *PME_Available_register* describes the capability of the PCS to aggregate multiple PMEs.[23] This register is read-only at the CO-side, and optionally writeable at the CPE-side (this should allow the subscriber to deliberately limit the aggregation possibilities). The *PME_Aggregate_register* selects the actual links to be aggregated into the same PCS. It is writeable at the CO-side, and remote-writeable at the CPE-side, by means of G.994.1 handshake messages.

The *Remote_Discovery_register* is there to allow the CO-side system to assess the interconnections between the PMEs and the PCS instances at the CPE-side of the link. Each PME can be actively connected to, at most, one PCS, although several PMEs can obviously be connected to the same PCS. The local system management uses a G.994.1 CL message, in response to the automatic CLR message from the CPE, to transmit a (preferably unique) 48-bit identifier[24] to the remote end, where it is to be stored in the *Remote_Discovery_register* of the PCS to which the receiving PME is connected.[25]

The CPE responds with another CLR message in which the actual value of the *Remote_Discovery_register* that was accessed is reported. This value will be equal to the value that was written, if and only if this

---

[23] The term PME is used here, as well as in the standard, to designate the entire part of the transceiver below the γ-interface, i.e., the combination of PMD, PMS-TC, and TPS-TC.

[24] The MAC address of the MAC to which the PCS is connected at the CO-side could be used as a unique 48-bit identifier.

[25] Extra signals are foreseen on the γ-interface at the CPE-side, to allow this value to be exchanged between the PME and the PCS.

First-time Startup

CO-side   CPE-side

C-TONES

R-TONE1

Discovery

10P/2B aggregation discovery code <= <MAC address>
Discovery operation <= 00_b

Normal handshake tones not shown

Discovery operation = 01_b

MR

REQ-CLR

CLR

CL

ACK(1)   Discovery

CLR

Startup if discovery
has previously taken place
or is not desired

CL

Tones exchange or keep-alive
sequence (silence message)
not shown

Aggregation

10P/2B link partner PME aggregate data <= <mapping>
Link partner aggregate operation <= 00_b

MR

REQ-CLR

Link partner aggregate operation = 01_b

CLR

CL

ACK(1)   Aggregation

CLR

CL

Startup if aggregation
has previously taken place
or is not desired

ACK(1)

Tones exchange or keep-alive
sequence (silence message)
not shown

Capabilities Exchange

MR

REQ-CLR

Link partner aggregate operation = 01_b

CLR

CL   Capabilities
Exchange

PMA/PMD link control = 1
Startup if capabilites exchange
has previously taken place
or is not desired

ACK(1)

Tones exchange or keep-alive
sequence (silence message)
not shown

Activation

MR

MS   Activation

ACK(1)

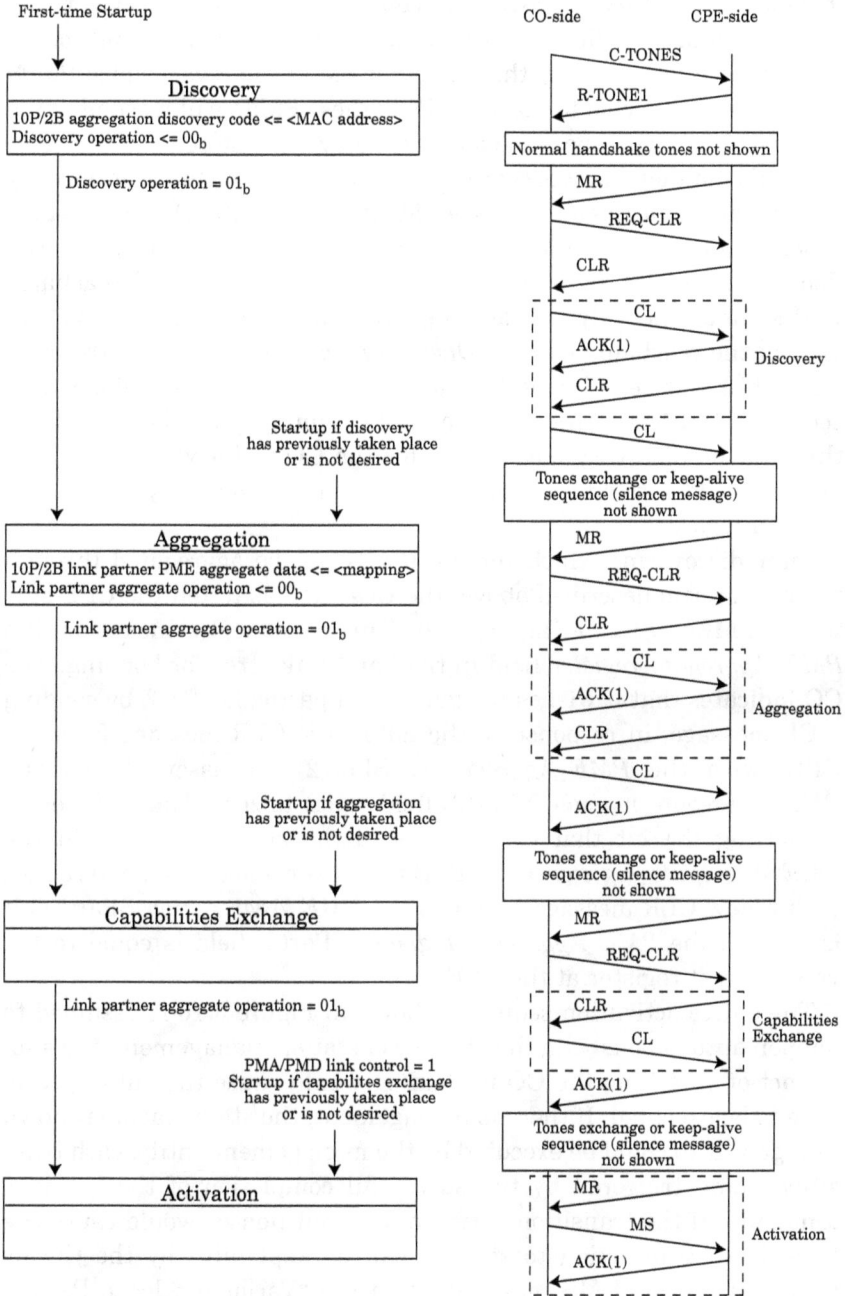

Figure 3.10   Link initialization state machine, including PAF discovery.

instance of the *Remote_Discovery_register* was not previously set to a
different value by a different "write" command. Subsequent reads of the
Remote_Discovery_register through other PMEs will reveal which PMEs
are actively connected to the same PCS; these PMEs will respond to the
read command with the same identifier previously stored.

In this manner, all aggregatable links can be brought up at once, by
sending different remote discovery identifiers to all PMEs at the same
time, or quickly one after another. The "write" transactions are specified to be atomic; therefore, even an attempt to write all reachable
PMEs simultaneously will actually write a single unique value into
each of the reachable *Remote_Discovery_registers*. Alternatively, after
a first link has been brought up with PME Aggregation enabled, subsequent PME Aggregation-enabled links coming up can be made to go
through the discovery mechanism for the CO to know whether they
are to be added to an existing aggregate, or whether they belong to a
new aggregate.

After discovering which remote PMEs *can* be aggregated through
the mechanism described above, the CO-side has to tell the CPE-side
which PMEs *shall* be aggregated. This is done by means of the
PME_Aggregate_register field in the handshake tree for bonding. The
CO indicates that it wishes to aggregate a particular PME by sending
a CL message, in response to the automatic CLR message from the
CPE, with the *PME_Aggregation* SPar(2) bit asserted and the
PME_Aggregate_register NPar(3) field equal to zero. The CPE reacts
by setting the bit that corresponds to the addressed PME in the
PME_Aggregate_register of the PCS that it is connected to, and replies
with a new CLR message in which the *PME_Aggregation* SPar(2) bit
is set and the *PME_Aggregate_register* NPar(3) field is equal to the
value of that register at the CPE.

The entire activation sequence shown in Figure 3.10 is assumed to
happen autonomously; in the absence of station management, the simple act of switching the CO-modem on would cause the entire procedure to be executed. If there is management, the state machine shown
in Figure 3.10 could be executed by the management entity, each transition being triggered by the successful completion of the previous
command. If the transition between different phases would cause the
handshake transaction to stall, it can be kept alive by the CO by
means of repeated MS messages with the "Variable Silence Period"
SPar(1) bit set.

The actual activation of the link happens by means of an MS message from the CO. If the capabilities unrelated to PME aggregation

were not exchanged during the discovery or aggregation phase, a third handshake phase is required prior to link activation.

Peer-to-Peer Handshaking with G.994.1

The EFM project is taking xDSL systems out of their familiar CO environment. EFM PHYs, especially 10PASS-TS, will be used in private networks (LANs) as well as in public networks. The context of a private network changes many of the assumptions under which the xDSL system will operate.

The G.994.1 handshaking procedure takes place between an HSTU-C (CO side) and an HSTU-R (subscriber side). In a private network, there is not necessarily any point that can naturally be labeled "Central Office," although this designation can be applied arbitrarily to one side of each individual link. In a tree-structured network, an n-port 10PASS-TS switch may act as a subscriber towards its uplink and as a CO towards its $n-1$ downlinks. Certain implementations may provide ports, which can be individually configured to operate as a CO-subtype or as a CPE-subtype. Getting a complex private network set up correctly requires a considerable amount of planning and configuration.

The previous Ethernet PHYs for twisted pair (10/100/1000BASE-T) share an optional auto-negotiation feature, which allows two stations on a link to decide which bitrate, modulation, and duplex mode will be used between them. A similar feature can be conceived for handshaking, to reduce the amount of manual configuration required in a private network.

The following algorithm can be used to autoconfigure a port that is capable of acting as a CO-subtype or a CPE-subtype:

1. Try to detect a C-TONES signal; if detected, assume the role of HSTU-R and proceed with handshake.
2. Try to detect an R-TONES-REQ signal; if detected, assume the role of HSTU-C and proceed with handshake.
3. Transmit an R-TONES-REQ signal. Try to detect a C-TONES signal; if detected, assume the role of HSTU-R and proceed with handshake.
4. Transmit a C-TONES signal. Try to detect an R-TONE1 signal or an R-FLAG1 signal; if detected, assume the role of HSTU-C and proceed with handshake.
5. If none of the above steps is successful, go into R-SILENT0 mode for a (random) amount of time, and return to step 1.

This algorithm will automatically lead to a correct configuration, irrespective of the nature of the transceiver on the other side of the link (CPE-subtype, CO-subtype, or autoconfigurable device). Step 4 may be omitted in devices that can be attached to a public network. This removes the possibility that a subscriber device assumes the role of a CO-subtype at its own initiative, which would cause interference issues.

Dual Latency and Pre-emption

Latency Paths in xDSL Standards

Unlike the EFM Copper PHYs, previous standard xDSL flavors provide two different latency paths at the γ-interface. The need for different paths comes from the different transmission requirements for certain applications or services.

An application transporting a real-time voice conversation over Ethernet (e.g., voice-over-IP) requires a very low latency in order for the subjective quality of the transmission to be acceptable for the human user. When the round-trip time exceeds about 400 ms, any amount of residual echo causes a doubletalk effect that makes it impossible for most people to carry on a conversation. Encoding and decoding of the speech signal will typically use up a significant part of that delay budget, so it's essential to keep the delay due to the transmission down to a minimum.

Prerecorded video content, on the other hand, does not require the same low latency (as long as it is one-way streaming). Even so-called "live" video content (e.g., televised news shows, sports events, concerts) does not lose its appeal under a transmission delay of up to several seconds (again, encoding and decoding of the video stream will take a significant amount of time). Loss of packets impacts the appreciation of the video quality much more; effects such as blocking artefacts and image freezes can be caused in theory by individual bit errors, leading to entire packets being discarded. Hence the need to keep the BER to a minimum, even at the expense of higher latency.

Dual latency deals with the voice/video dichotomy by providing a programmable interleaver and a way to bypass it.

- The *slow path* or *interleaved path* is optimized for video: It sends the Reed-Solomon encoded data stream through a programmable interleaver, to make it more robust against impulse noise. Impulse noise hitting the interleaved bit stream will be smeared out over several Reed-Solomon frames after de-interleaving, improving the chance of a complete error correction. This mechanism lowers the BER at the expense of up to 20 ms of latency.

- The *fast path* or *non-interleaved path* is optimized for voice: The interleaver is bypassed, obtaining the lowest possible latency at the expense of less robustness against impulse noise.

In ADSL and VDSL standards, only the slow path is mandatory. If the interleaver is programmable, it can be set to a very low (or zero) depth, generating the same effect as a fast path. Support for both paths, that is, dual latency, is optional. During initialization, the CO-side transceiver (configured by the operator) will request certain upstream and downstream bit rates for the slow path and optionally the fast path.

Queuing Issues and Pre-emption

Ethernet frames vary in size between 64 bytes and 1522 bytes (not counting the 7 bytes of preamble and the single-byte start-of-frame delimiter). At a typical ADSL speed of 1 Mbps, a maximum-size Ethernet frame takes 12 ms for transmission. When different services are mixed over a common Ethernet-over-xDSL link, packets with different priorities will be queued in a single queue at the obvious bottleneck: just above the xDSL transmitter. The more urgent frames (e.g., voice-over-IP packets) are typically short, whereas the time-insensitive HTTP and FTP packets tend to be more bulky. A situation that often arises is when an urgent little voice packet is queued after a big slow web packet....

In ATM, the queuing problem is solved by prefragmenting all data into 53-byte cells. There will still be a need for queuing, but the amount of lost time waiting for transmission of a lower-priority packet is limited to 53 bytes at transmission rate. Fragmentation to improve queuing characteristics has been proposed in native packet-based protocols, such as the Point-to-Point Protocol (RFC2686 and RFC2687).

In the absence of such a fragmentation mechanism for Ethernet, it can be shown that performance can be improved by dividing the bandwidth in different paths and providing independent queues for these paths, even when the interleaver depth of both paths is identical [7]. If Ethernet is to be used over low-speed xDSL links, the concept of "paths" will be useful to improve multi-service performance. Dual latency offers the possibility of having two logical paths on the same xDSL link, but it requires provisioning of a fixed amount of bandwidth for each of the paths.

Pre-emption, based on a suspend/resume mechanism, offers the benefits of multiple logical paths without the need to provision fixed amounts of bandwidth for each path. Any data offered at the interface of the pre-emptive path will cause the ongoing transmission of non–pre-emptive data (if there are any) to be suspended for as long as it takes to transmit the pre-emptive data. On the receiver side, pre-emptive data and non–pre-emptive data are demultiplexed and presented to their respective interfaces. The TC sublayer mechanisms required for pre-emption were explained in the section "64/65-Oclet Encapsulation."

Dealing with Multiple Paths in Ethernet

In ATM-based systems, the γ-interface is typically implemented as a Utopia-interface. Due to the bus-like nature of Utopia, this eliminates the need for two physical interfaces; it is sufficient to have different addresses for the slow path and the fast path.

In Ethernet, the interface between a host (or its MAC) and a PHY is point-to-point. A PHY implementing dual latency or pre-emption would have to present two MIIs to the outside world and thus be considered as two PHYs—a PHY implementing both dual latency and pre-emption would even need four MIIs (see Figure 3.11)! If there were only one MII on a dual-latency PHY, there is no way to indicate which path is appropriate for a given frame. This is the main reason why neither dual latency nor pre-emption were selected as a part of the EFM standard.

Let us have a look at what could be done to accommodate a *non-standard* multiple-path PHY in an Ethernet system. Without loss of generality, we will discuss the case of two paths. The case for more than two paths is analogous.

Separate LANs. A multi-path Ethernet-over-xDSL implementation with two MIIs can simultaneously offer connectivity to two classes of hosts. At the customer's premises, a different bridged LAN could be attached to each of the paths, the first one grouping all latency-sensitive equipment (IP phones, gaming stations, etc.), the second one containing the latency-insensitive ones (video set-top box, TV browser, etc.).

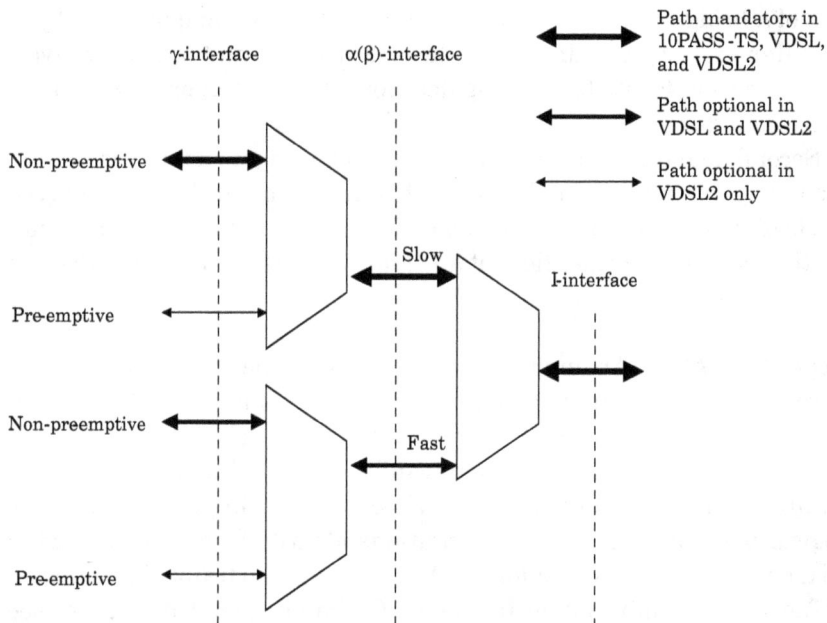

Figure 3.11   Multiplexing of different logical paths in VDSL-based PHYs.

To this end, the modem should have (at least) two Ethernet ports, which could have different colors to be easily recognizable. With the exception of the common Ethernet-over-xDSL transceiver, the two LANs must be physically separate. The advantage is that the physical network architecture reflects the separation between the two classes of services. The obvious disadvantage is the need to purchase and maintain two sets of bridged LAN equipment (bridges and wiring).

Integration in forwarding fabric.   It may seem tempting to connect the different MIIs of the multi-path xDSL transceiver directly to a MAC bridge and let the bridge take care of selecting the most appropriate port for every frame received from the LAN according to traditional bridging rules. There are two serious drawbacks to this solution.

First, this solution may only be used at one end of the link. If both ends of the link tie up the different logical paths to a bridge, a loop is created between the bridges that will cause one of the logical paths to be disabled by the spanning tree protocol running on the bridges (see Chapter 10)—this obviously defeats the purpose of having different logical paths. As a result, at least one side of the link must connect the different logical paths to different LANs. This does not preclude the use of a bridge on one end and a router on the other end of the link, as a router properly terminates the layer-2 network and uses only layer-3 information for its forwarding decisions, but that approach requires more configuration.

Second, leaving the selection of a logical path up to a forwarding fabric (bridge or router) implies that this selection will be made on the basis of a destination address alone. The destination address may not be the most appropriate field of a frame on which to base the selection of a logical path.

Separate VLANs. An improvement over the separate LAN architecture, with the benefit of being an "integrated" solution like the bridge-based solution described earlier, is the use of virtual bridged LANs (VLANs). All customer equipment can be connected to a single VLAN-aware bridge, on which different VLANs are defined for each of the logical paths on the xDSL link. Stations identify frames as belonging to a particular VLAN by adding a VLAN tag to the frame (indicated as QTag Control Information in the MAC Frame format definition, see Table 1.1). A VLAN-aware bridge has independent forwarding tables for frames belonging to different VLANs, and hence presents the behavior of different independent bridges in one box (see the section "VLAN Bridges" in Chapter 10 for further details).

The advantage of this approach is the use of common network infrastructure for two virtual networks. The disadvantage of using VLANs is the need for VLAN-aware equipment at the customer's premises.

Fast/slow aggregation. A final way to deal with multi-path–capable Ethernet equipment requires only one LAN infrastructure. The two paths are aggregated so that layers above the path selection logic see the whole as a single link, much the same way as link aggregation aggregates identical Ethernet links into a fat Ethernet pipe. This is done at both sides of the link.

The mechanism is based on the concept of *conversations*: These are streams of Ethernet frames sharing a common source address *and* destination address. Whereas link aggregation is used to provide bit rates, which are integer multiples of existing Ethernet links (e.g., $n \times$ 100 Mbps), or to implement redundant links, fast/slow aggregation divides the stream of datagrams coming in from the MAC Client into conversations, which are mapped on the low-latency path or the high-latency path of Ethernet-over-xDSL link. Remembering that a conversation is defined by its source and destination address, it provides some information about the nature of the communication. If the source address of the frame is the customer's IP phone, and the destination address designates a remote voice-over-IP gateway, this information should clearly be sufficient to assign that conversation to the fast path. In an analogous way, a number of other services can be identified by source and destination address. The remaining conversations can be assigned to a default path.

By implementing this mechanism and a user-configurable table (associating pairs of MAC addresses with services) in a multi-path Ethernet-over-xDSL PHY system, this system may now present a single MAC service interface to its client (see Figure 3.12). In this case, as in link aggregation, the MAC service interface really hides an aggregation sublayer plus several underlying MACs and virtual PHYs. Although additional user-configuration is necessary to make this approach work, it effectively removes the disadvantages of the methods described earlier.

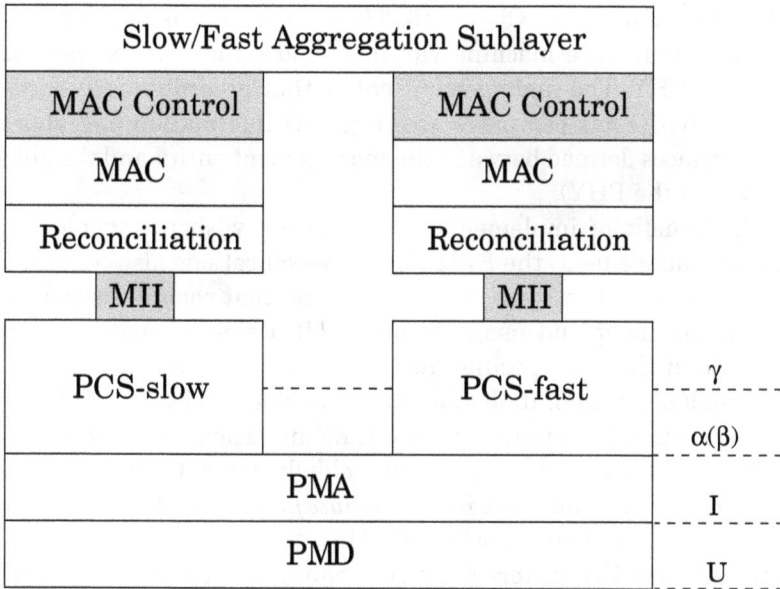

Slow/Fast Aggregation Sublayer		
MAC Control		MAC Control
MAC		MAC
Reconciliation		Reconciliation
MII		MII
PCS-slow	- - - - -	PCS-fast
PMA		
PMD		

Figure 3.12   Aggregation of different latency paths in EFM Copper.

There is no need to limit the path selection logic of the fast/slow aggregation sublayer to selections based on the conversation to which frames belong. Although there is no precedent for it in the standard, a valid alternative would be to select paths on the basis of the user priority bits, if present. If any other selection criteria are used, the in-order delivery of frames belonging to a certain conversation and a certain user priority can no longer be guaranteed.

Management for EFM Copper

The Ethernet standard specifies two interfaces for access to the registers that contain the PHY's management information. Clause 22 contains the specification of the MII (also see the section "Operation of MAC and MII" in Chapter 2), including an optional serial interface for station management (STA). Clause 45 specifies a management data input/output (MDIO), also optional, which is an extension to the MII management interface, originally intended for transceivers operating at 10 Gbps or above. It provides access to more registers while retaining logical compatibility with the original MII management interface.

EFM Copper provides complete access to all features of the PHYs by means of the Clause 45 MDIO. A large set of dedicated registers was added for this purpose to Clause 45. These registers correspond to or are derived from state machine variables and signals in the various parts of the PHY. The management entity that presents these registers to the MDIO has pervasive access to this information (i.e., there are no interfaces defined between the management entity and the sublayers within the PHY).

For the benefit of implementers who do not wish to include full Clause 45 management, the EFM Copper specifications also contain a number of profiles; these are "typical settings" that should cover most real-life deployment and usage scenarios. If a simpler management interface than the one specified in Clause 45 is present in an EFM Copper implementation, it should allow users or operators to select a mode of operation from these profiles. If no management is present at all—which is possible, but certainly unlikely for a CO-type transceiver—the PHY should start in the *default profile* (see Chapter 4 for 10PASS-TS and Chapter 5 for 2BASE-TL).

Certain Clause 45 registers represent management information pertaining to the remote PHY. This information is exchanged either during initialization, by means of G.994.1 handshaking, or during data mode. Each of the EFM Copper PHYs has certain out-of-band management channels, which are active during data mode for the purpose of management (see Chapters 4 and 5 for PHY-specific details).

Note that the G.994.1 codepoint trees define a format for handshaking *messages*, they do not constitute an independent set of *variables* by themselves. The handshaking messages merely convey the value of certain constants and variables that are internal to the PHY, some of which may be accessible through management.

The management functions provided by the OAM sublayer, as described in Chapter 8, do not cover any functions that are required for normal operation.

The EFM OAM protocol or any other higher-layer management protocol may be used to access those resources belonging to the remote PHY that are not locally available through Clause 45 registers. The MIB in Clause 30 of IEEE Std. 802.3 may be used with the EFM OAM protocol; MIBs for use with SNMPv2 are currently being defined in the IETF.

4

The Short-Reach Solution: 10PASS-TS

"We have two firmly entrenched camps that have been battling for nearly a decade over the choice between QAM (aka SCM or single carrier modulation) and DMT (aka MCM or multi-carrier modulation). This is a religious war, with fortunes to be made or lost on each side. As usual, the primary battle lines have been drawn by the IC vendors, but there are zealots among the equipment vendors and service providers as well."

—HOWARD M. FRAZIER

Introduction to VDSL

General Characteristics

Of all standardized xDSL flavors, Very-High Speed Digital Subscriber Line (VDSL) offers the highest bitrates. To accomplish this, VDSL transceivers use a much wider spectrum than any other xDSL: Standards originally provided for bandwidths up to 30 MHz, but currently approved frequency plans are all limited to 12 MHz (see Figure 4.3)—this is already a 10-fold increase over the 1.1 MHz of bandwidth used by ADSL.

As is explained in Chapter 3, signal attenuation increases with frequency and cable length, so the higher frequency range can only be exploited on very short loops. A typical reach often quoted for VDSL is 1.5 km, which limits its deployment options to private networks (e.g., LAN extentions) and phone lines served from a street cabinet. The high investment associated with building fiber-fed cabinets for VDSL distribution is probably the main reason for the slow take-up rate in most parts of the world.

Single-Carrier Modulation

Quadrature Amplitude Modulation (QAM) is a modulation method in which information is coded in the amplitude and the phase of a carrier wave with frequency f. A QAM symbol can be written as

$$s_{m,n}(t) = A_m \cos(2\pi\, ft) + B_n \cos(2\pi\, ft) \tag{4.1}$$

which is equivalent to the real part of

$$s_{m,n}(t) = (A_m - jB_n)e^{j2\pi\, ft} \tag{4.2}$$

The former notation explains the term "quadrature" in the name: the symbol is the superposition of two waves of the same frequency, shifted in phase by a quadrant (i.e., the phase difference is $\pi/2$). The latter notation shows that the result of this superposition is another wave of the same frequency, shifted in phase.

The carrier amplitudes A_m and B_n are chosen from a discrete set[1] called a *constellation*, often represented in a two-dimensional grid, the size of which depends on the number of bits to be transmitted per symbol (see example in Figure 4.1). Each point in the chosen QAM constellation corresponds to a predefined bit sequence.

After appropriate equalization to compensate for the frequency-dependent attenuation and phase rotation introduced by the channel, the received symbols are demodulated (this corresponds mathematically to a discrete-time projection of the received symbol onto an unmodulated carrier), and the bit sequence can be regenerated.

However, the influence of noise (including RFI and crosstalk) cannot be removed by equalization. Noise will have the effect of moving the demodulated constellation point away from its target position on the grid. Let us assume for a moment that the average energy of the received constellation points can be chosen to be any reference value through equalization and amplification. The higher the order of the constellation, the denser the grid of constellation points becomes. The average drift of the demodulated constellation point away from its

---

[1] This is *digital* communications after all! Replace the factors A_m and B_n by two functions $A(t)$ and $B(t)$, and you get the well known analog version of QAM, used in radio frequency communications.

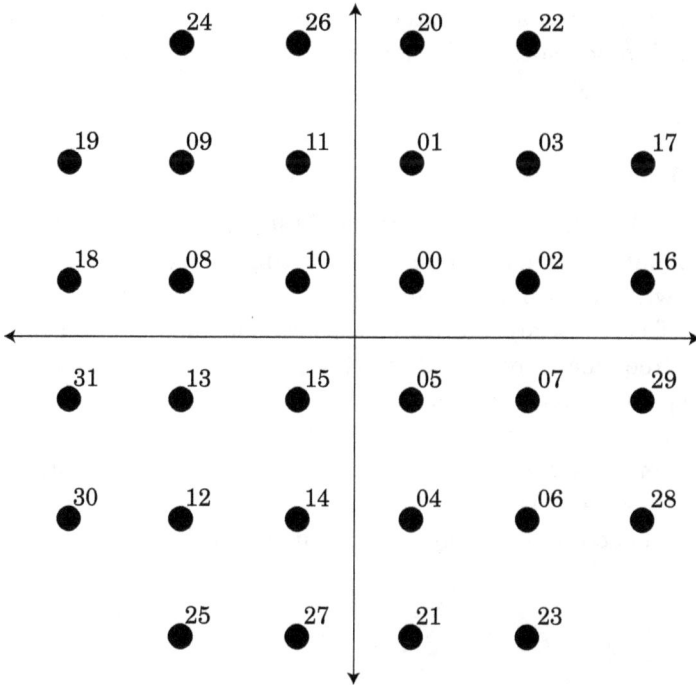

Figure 4.1   Constellation labels for $b=5$ (32-QAM).

original position on the grid is proportional to the amount of noise power in the received signal after equalization. The maximum density of the QAM constellation one can use and hope to demodulate successfully on a given channel thus depends on the signal-to-noise ratio (SNR). With certain simplifying assumptions about the nature of the noise, the minimum SNR needed to obtain a certain bit error ratio can be calculated for each constellation size.

Single-carrier VDSL, also known as *SCM* or simply as *QAM*, uses a single carrier per frequency band, modulated in QAM. The symbol rates are chosen in function of the spectral width of each of the bands. Each carrier carries up to 8 bits per symbol (12 in the version proposed for Ethernet in the First Mile), corresponding to a 256-QAM constellation (4096-QAM in the version proposed for EFM). Due to the high spectral width of the carriers, sophisticated equalization is needed in the receiver.

Early implementations of QAM-VDSL used only one carrier in each direction, thus further simplifying the digital processing. These systems would typically yield bit rates up to about 15 Mbps in each direction, and were marketed as an extended-reach version of 10BASE-T Ethernet (with names such as V-thernet, 10BaseS, and Long Reach Ethernet).[2]

Discrete Multi-Tone

Modulation.   In discrete multi-tone (DMT) VDSL, also known as multi-carrier VDSL, data is transmitted by modulating a large set of carriers in QAM, with an equal frequency spacing of 4.3125 kHz.[3] Two disjoint sets of carriers are used for transmission in upstream and downstream (frequency division duplexing), according to regionally regulated frequency plans, spanning a bandwidth of up to 12 MHz. These frequency plans are further discussed later in this chapter.

Mapping a set of N complex QAM constellation points F_k onto N equally spaced carriers is exactly what an inverse discrete fourier transform (IDFT) accomplishes by virtue of its definition:

$$f_n = \frac{1}{\sqrt{N}} \sum_{k=0}^{N-1} F_k e^{j2\pi nk\,/\,N} \qquad (4.3)$$

Mapping a set of QAM constellation points onto equally spaced carriers is exactly what an inverse discrete fourier transform (IDFT) accomplishes. The additional advantage one obtains when using IDFT as a modulation technique is that all the transmitted carriers are mutually orthogonal. DMT is thus a form of orthogonal frequency domain multiplexing (OFDM), where different carriers may carry a different number of bits per symbol. Moreover, the carriers used in each direction are also mutually orthogonal, which is exploited under the form of "digital duplexing" (see Figure 4.2, where the role of the cyclic extensions is also illustrated).

The well-known correspondence between multiplication in the time domain and convolution in the frequency domain can help us understand the importance of orthogonality (and of timing) in DMT systems. In the time domain, anything that deviates from a strictly repetitive sequence of samples with a frequency that divides the sampling fre-

---

[2] Here, "long reach" is to be interpreted relative to the 100 m limitation of the 10BASE-T LAN technology. In a subscriber access network, VDSL is considered to be a "short reach" technology.

[3] An alternative tone spacing of 8.625kHz is supported by some implementations, but supporting this mode is not mandatory in any standard.

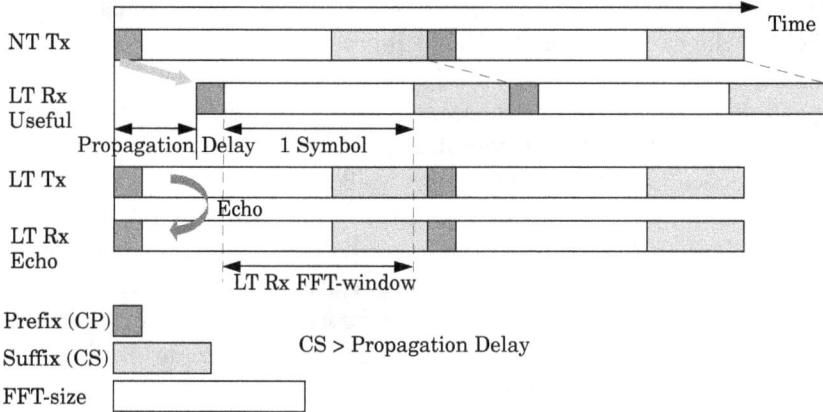

Figure 4.2   Digital duplexing as applied in DMT-VDSL Source: EFM Task Force [5].

quency, will turn out in the frequency domain as power dispersed over multiple carriers. Obviously, DMT symbols have a finite duration, and are thus not a repetitive sequence.

If the symbol sampling window at the receiver was to include a point in time t_c at which a first transmitted symbol ends and a second transmitted symbol starts, the set of samples could be interpreted as the sum of the first symbol (assumed to be forever repeating in time) multiplied with a step function, which is non-zero for $t < t_c$ and zero for $t \geq t_c$, and the second symbol multiplied with a step function, which is zero for $t < t_c$ and non-zero for $t \geq t_c$. In the frequency domain, this will not only cause the amplitudes for each carrier in the two symbols to be unresolvably mixed, but the carriers will be convoluted with the Fourier transform of the step functions (this effect will manifest itself as sidelobes).[4]

From this discussion, the need for synchronization between transmitter and receiver will seem obvious. However, we can go further. Orthogonality may also be achieved between different systems. Indeed, when different VDSL systems sharing the same cable bundle synchronize their symbol transmissions, the near-end crosstalk (NEXT), which is mostly caused by sidelobes from the transmit carriers that are picked up in the receive spectrum, is made orthogonal to the received signal, which effectively eliminates the NEXT. This NEXT suppression property may be used between DMT-VDSL systems and even ADSL systems using the 4.3125-kHz tone spacing. Different tone spacings (e.g., 8.625

---

[4] A practical calculation of the sidelobes of a DMT transmitter and a lot of other detailed information about DMT technology can be found in [8].

kHz) imply different symbol periods, so tone spacings must not be mixed in the same binder if NEXT suppression is desired.

Equalization.   Each carrier represents only a very small portion of the bandwidth available to the complete transceiver system, so the channel properties (transfer function, crosstalk, noise) vary only very little within the bandwidth of a single carrier. The result is that the channel equalization to be performed for each carrier is greatly simplified with respect to SCM. Equalization is usually achieved by combining a time-domain equalizer (finite impulse response filter applied to the received signal) and a frequency-domain complex multiplication (phase and amplitude compensation per tone).

To suppress inter-symbol interference, caused by symbol dispersion during transmission over the wire, a cyclic extension is appended to the symbol. This extension serves as a buffer between adjacent symbols. Another purpose of the cyclic extension is digital duplexing, consisting of the alignment of the receiver window with the transmit symbols in such a way that no transmit symbol boundary occurs during the time in which a single receive symbol is read (shown in Figure 4.2). This alignment ensures orthogonality between transmit signal and receive signal, thus effectively eliminating echo of the transmit signal into adjacent receive carriers.

Bit loading.   Given the SNR of each carrier, the desired bit error ratio (BER), and the desired noise margin, a number of bits b_i are "loaded" onto each carrier ($b_i \in \{0, ..., 15\}$). This means that for each carrier, an appropriate 2^{b_i}-QAM constellation is selected that offers the requested BER or better, so that the sum of all b_i over all available carriers multiplied by the symbol rate (4000 symbols per second) is equal to the requested bit rate. The bit loading is under control of the receiver, as the SNR can only be measured precisely at the receiver side. Rate adaptive operations allows a VDSL system to retrain to a lower bit rate when the constraints on the BER and noise margin can no longer be met due to a decreasing SNR.

Gain scaling is introduced to further optimize the bit loading algorithm. Carriers that lack only a little SNR to support an additional bit (i.e., the next QAM constellation) may be boosted by up to 2.5 dB on request of the receiver, at the expense of other carriers that have up to 2.5 dB of SNR to spare. This results in a set of gain scaling factors g_i. The algorithms to select b_i and g_i values are not standardized.

The selected b_i and g_i values are transmitted back to the link partner during initialization. To do this in an efficient manner, b_i and g_i are specified for groups of tones, and each group may be specified by the values at $j_{max}+1$ carriers, where a polynomial of order j_{max} is used to interpolate between these carriers.

Framing. Payload data and overhead channels are multiplexed into a single octet stream in the PMS-TC sublayer. At this level, a "frame" is a set of bytes carried by one DMT symbol. Each frame carries bytes belonging to each of the data paths (the fast path is optional) and the overhead channels. The amount of slow path or fast path data carried by a single frame is called a "packet." Processing of these TPS-TC packets consists of the following steps:

Payload traffic adaptation. Packets from each of the latency paths are grouped into sets of 138 packets each to achieve "payload traffic adaptation." This concept allows bitrates that are integer multiples of 64 kbps to exist at the $\alpha(\beta)$-interface, with the various lengths of the cyclic extensions. This is accomplished by adding an appropriate number of dummy bytes in function of the selected cyclic extension length, to each set of 138 packets. This implies that a small fraction of the capacity of the line is given up to dummy bytes in order to reach the 64 kbps granularity at the $\alpha(\beta)$-interface.

Fast or slow octet insertion. These octets carry the overhead channels, as well as a sync byte, a CRC, and a Network Timing Reference. The overhead channels are the embedded operational channel (EOC), VDSL Overhead Channel (VOC), and Indicator Bits (IBs). They carry system-related, path-related, and line-related information across the line, which is necessary for normal operation. This exchange of information must not be confused with the EFM OAM frames described in Chapter 8.

Reed-Solomon dummies insertion and encoding. Reed-Solomon coding, which is described in detail later in this chapter, transforms a set of K octets, into an N octet codeword. A number of dummy bytes are added to the data in order to have an integer number of RS-codewords in each set of N packets.

Fast and slow buffer merge. Finally, the fast buffer and the slow buffer, each containing RS-encoded payload data and overhead bytes, are merged into a frame, which can be mapped onto a single DMT-symbol.

To find out which line rate is required to support a given payload rate at the α(β)-interface, one has to add traffic adaptation dummy bytes, Fast/Slow octets, RS dummies, and RS coding overhead.

Strengths and weaknesses. The downside of the DMT transmission technique is the fact that calculating an (inverse) discrete Fourier transform requires significant computing power. However, this complexity is purely in the digital domain, which easily scales down with every new step in silicon integration technology; the simplicity gained in the analog signal processing will soon outweigh the digital disadvantages.

The main strength of DMT lies in its ability to adapt to conditions in which the SNR is not a smooth function of frequency, e.g., in the presence of bridged taps or radio-frequency interferers. Conversely, the transmit power spectral density can be tuned with 4.3125-kHz granularity, allowing for fine notches in the emission spectrum, which may be required to protect AM broadcast stations or HAM radio bands.

From the point of view of deployment, DMT has the advantage of being able to implement any desired frequency plan simply by a change in the allocation of carriers to the upstream or downstream, which can be done from the management software. Apart from being able to obey all of the regionally mandatory frequency plans, this allows for any desired symmetry ratio when used in privately owned networks.[5]

VDSL Standards

In the past three years, four different SDOs have been developing or enhancing VDSL standards. The result is that there are four standards of the first VDSL generation (hereafter called VDSL1). Two Standards Development Organizations (SDOs) have selected multi-carrier modulation (T1E1.4 and IEEE 802.3ah); the other two decided to standardize both multi-carrier and single-carrier modulation (not mutually interoperable).

A new generation of VDSL standards, the "VDSL2" generation, is expected to count at least three members, all using multi-carrier mod-

---

[5] In an early implementation known as the "Zipper," carriers were alternatingly assigned to upstream and downstream, yielding a perfectly symmetrical bit rate at any loop length. However, such systems need to be synchronized with other Zipper systems using the same binder, to ensure orthogonality between systems (otherwise the NEXT would seriously degrade performance). This is a very unpractical solution in an unbundled environment.

ulation. The goal of VDSL2 is to upgrade VDSL1 with features from the more recent ADSL2 standard, to get the "best of both worlds." VDSL2 will provide better bitrates and reaches, support native packet transport (see also "The EFM-TC in Other Standards" in Chapter 3), and generally simplify the design of a combined ADSL2/VDSL2 transceiver. Among the performance-improving features of VDSL2 are the support for constellations up to 15 bits, trellis coding, and an increase of the bandwidth up to 30 MHz.

ETSI TM6. The European Telecom Standardization Institute (ETSI) was the first SDO to complete a VDSL standard. ETSI TS 101 270 is a standard in two parts. The first part contains the functional requirements for VDSL transmission systems. The second part specifies two transceivers: Both multi-carrier VDSL and single-carrier VDSL are completely standardized, and the choice is left to the implementer. The ETSI VDSL standard is intended for use in Europe only. It supports two frequency plans: Plan A (mandatory) and Plan B (optional). The document contains a TPS-TC for ATM transport and one for STM transport.

ETSI TM6 is currently working on a VDSL2 specification, which is to use multi-carrier modulation only.

ATIS NIPP-NAI (formerly T1E1.4). The American National Standards Institute (ANSI)-accredited Committee T1E1.4 published a trial-use standard for VDSL during the development of the EFM standard. T1.424/Trial-Use was a standard in three parts. The first part contained the functional requirements and common specifications for VDSL interfaces. The second part specified a transceiver based on single-carrier modulation. The third part specified a transceiver based on multi-carrier modulation. The choice between part 2 and part 3 was left to the implementer. During the two-year validity of the trial-use standard, the linecode war was fought (see below), which eventually resulted in the exclusive selection of multi-carrier modulation for the American National Standard T1.424 (the official successor of T1.424/Trial-Use). The single-carrier VDSL specification is republished and maintained as a technical requirements (TRQ) document; this is a technical publication without the "standard" status. Both documents are intended for use in North America only; plan A is the only supported frequency plan.

American National Standard T1.424 and the single-carrier modulation TRQ contain a TPS-TC for ATM transport and one for STM transport. The new ANSI standard on "Ethernet Bonding" contains normative references to the EFM-TC specified in IEEE 802.3ah, but the EFM-TC can only be used by transceivers that have a capability to select this TPS-TC in their G.994.1 handshake codepoint tree (which excludes both the American National Standard and the TRQ).

ATIS NIPP-NAI is currently working on a VDSL2 specification, which is to use multi-carrier modulation only.

ITU-T Q4/15. For a long time, *Question 4* of ITU-T Study Group 15 had no other VDSL specification than the "foundation document" that was published as Recommendation G.993.1 (2001). This document avoided the linecode question by not specifying the Physical Medium Dependent Sublayer (PMD). Q4/15 has recently published a revision of this Recommendation, which specifies multi-carrier modulation in the main body of the document, and single-carrier modulation in a normative annex.[6] ITU-T Recommendation G.993.1 is intended for international use. Different Annexes specify the operational parameters for different geographical regions; it is not necessary to support operation in all regions to claim G.993.1 compliance. G.993.1 contains TPS-TCs for ATM, STM, and PTM transport.

The revised G.993.1 document contains a "death clause" for QAM: a statement that future revisions of the Recommendation (VDSL2) will be based on DMT only.

IEEE 802.3ah. The 10PASS-TS standard is specified as a reference to American National Standard T1.424, with a list of exceptions and additions. Just like T1.424, it uses multi-carrier modulation. Unlike T1.424, it contains provisions for international use; not only are all ITU-T frequency plans explicitly supported, but there is also the flexibility to define new frequency plans for use in private networks. All requirements must be supported by the system to claim 10PASS-TS compliance.

There is currently no intention in the IEEE 802.3 Working Group to start a project to add a VDSL2-based Physical Layer entity sublayer (PHY) to the Ethernet standard. The VDSL2 standards under development in the other SDOs will include 64/65-octet encapsulation as a "packet mode" and have access to Ethernet bonding through new standards pointing to the EFM PCS, so VDSL2 will be the *de facto* successor of 10PASS-TS.

---

[6] Read "Linecode Wars and Tone Spacing" to see how this agreement came about.

With so many different VDSL standards to choose from, it is justified to ask what the role of the 10PASS-TS specification in IEEE Std. 802.3ah will be. Detailed drafts of VDSL1 standards have been around much longer than the 10PASS-TS specification, and VDSL1 is currently being deployed. 10PASS-TS is viewed by many as "the same thing as VDSL1 with 64/65-octet encapsulation and bonding added" (it is one of the goals of this book to eradicate that view!). As an understandable result, the first generation of so-called 10PASS-TS modems are improved VDSL1 modems with some of the key 10PASS-TS features added, but without full compliance to the letter of the standard (specifically, without the internationalization features discussed in the section "Profiles and Internationalization"). These modems have to compete with the installed base of HDLC-based VDSL1 modems.

Meanwhile, VDSL2 is progressing, and it is likely to be standardized by the time this book is published. Modem builders are already building in prestandard VDSL2 features, as customers are already asking for them. It is possible that VDSL1 may be supplanted by its feature-rich successor VDSL2, without leaving a market window for 10PASS-TS, but that is impossible to predict.

Frequency Plans

ITU-T Recommendation G.993.1 specifies the three frequency plans shown in Figure 4.3. Each plan consists of an optional band (25—138 kHz), to be used in upstream or downstream, followed by a first downstream band (D1), a first upstream band (U1), a second downstream band (D2), and a second upstream band (U2).

- **Plan A** (in Annex A), formerly known as Plan 998, is used in North America. [Its use is prescribed by the American VDSL standard (T1.424), and by spectrum management rules (American National Standard T1.417].
- **Plan B** (in Annex B), formerly known as Plan 997, is used in some European countries. It is much better suited for symmetric services than Plan A.
- **Plan C** (in Annex C), formerly known as Plan Fx, is used in Sweden. The split frequency between the second downstream band and the second upstream band is a variable, F_x, which can be adjusted to provide the desired ratio between downstream bitrate and upstream bitrate at a given reach.

Annex A

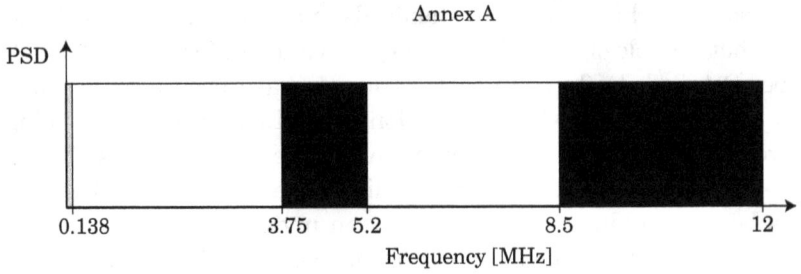

Annex B

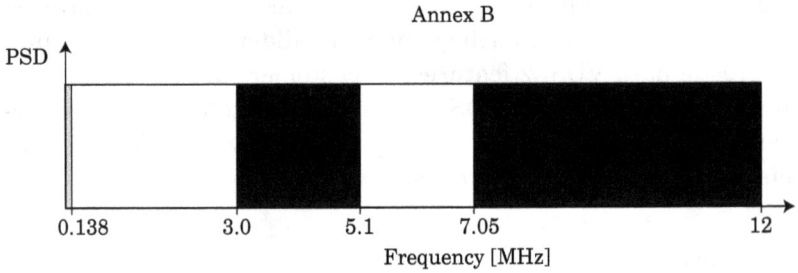

Annex C

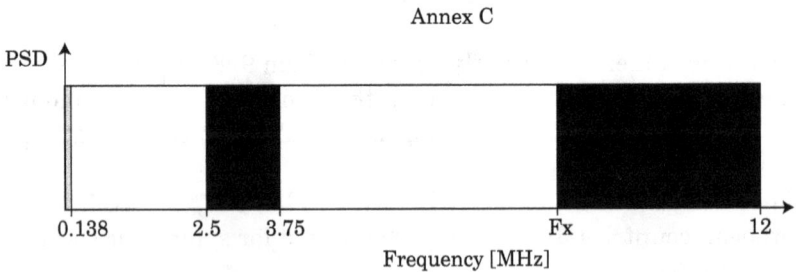

Figure 4.3   VDSL frequency plans, as defined in ITU-T Recommendation G.993.1

Upstream Power Back-Off

The fact that VDSL uses frequencies that are much higher than those used by other xDSL flavors makes it particularly vulnerable to crosstalk (see Equation 3.1 and Equation 3.2 for the frequency dependence of crosstalk models). It is possible two VDSL lines in a cable bundle differ so much in length that the crosstalk generated by the nearest customer premises equipment (CPE) is stronger than the upstream signal received from the furthest CPE. In such a situation, the subscriber on the longest line would not be able to set up a link.

To resolve this problem, upstream power back-off (UPBO) was introduced. The principle consists of reducing the upstream transmit power of all CPEs to achieve the same "reference PSD" at the Central Office (CO). For this purpose, the CO modem transmits the selected reference PSD to the CPE during initialization, and the CPE then reduces its output power based on this reference PSD and the estimated loop length. If necessary, the CO modem can then instruct the CPE to further fine-tune its transmit PSD.

An additional mechanism available to DMT-VDSL (and hence, 10PASS-TS) systems, is the specification of a maximum upstream transmit PSD by the CO. This maximum upstream transmit PSD may further restrict the PSD used by the CPE after application of the reference PSD method. It must not cause the CPE to transmit a PSD that would exceed the one obtained by the reference PSD method.

Although it may be possible to turn UPBO off for testing purposes (this is indeed required for certain tests specified in the 10PASS-TS standard), UPBO should always be enabled in real deployments. G.994.1 handshaking and part of the initialization sequence occur before UPBO is applied and may cause transient disturbance to other VDSL systems in a binder.

Forward Error Correction

Error detection, without the capability to correct detected errors, may be useful in situations where loss of data is tolerable, for example, either because the receiver can do without the data, or because it has a way of requesting retransmission of the missing data. The IP protocol, Ethernet's favorite passenger, is tolerant to loss of data, because its transport protocols UDP and TCP are loss-insensitive and capable of retransmission, respectively. This is why Ethernet's error-detection, achieved by a 32-bit CRC, followed by frame discard, is perfectly acceptable to IP.

At the bottom of the physical layer (PMD and PMA), different rules apply. From α-interface to β-interface, the xDSL system behaves as a byte pump without any frame-awareness. This lack of frame-awareness, combined with the low latency and BER requirements at this level, makes mere error detection unsuitable. Forward Error Correction (FEC) is used to detect and correct errors and is particularly useful when there is no way to request retransmission of the corrupted data, which is why it is used in modems and storage systems.

VDSL and ADSL systems use a Reed-Solomon code for FEC, operating on GF(256). A Galois field $GF(n^m)$ is a *finite* field, consisting of a set of n^m elements that can be written as consecutive powers of a primitive element α.

$$\{0,\alpha^0,\alpha^1,\alpha^2,...\} \tag{4.4}$$

It is a *field* because addition and multiplication operations are defined over the set, and these operations have specific properties. The set is made finite and closed under multiplication by adding a closing constraint.

$$\alpha^{n^{m-1}} = \alpha \tag{4.5}$$

GF(256), or $GF(2^8)$, is the Galois field with 256 elements, used as "symbols" to construct RS codewords. The field elements are generated by a *primitive element* α, which is a solution of an equation known as the *primitive polynomial* of the field.

$$\alpha^8 + \alpha^4 + \alpha^3 + \alpha^2 + 1 = 0 \tag{4.6}$$

$$\Rightarrow \alpha^8 + \alpha^4 + \alpha^3 + \alpha^2 + 1 \tag{4.7}$$

This equation imposes a certain arithmetic on the field, by implying that every non-zero element of GF(256) can be written as a weighted sum of the elements $\{\alpha^0,\alpha^1,\alpha^2,...\alpha^7\}$. By using the bits of a single octet as the coefficients for the values $\{\alpha^0,\alpha^1,\alpha^2,...\alpha^7\}$, we can map any octet on a unique member of GF(256). This clearly shows that GF(256) is perfectly suited to represent the set of all possible octet values, which is why it is popular in data communication applications. With this mapping of octets to GF(256) elements, we now have a way to add and multiply octets, which is all we need to get to the next step: polynomial division.

What happens next is analogous to CRC calculation; only, the operations are performed in GF(256) instead of in the binary field. A Reed-

Solomon code word contains $N=K+R$ bytes, where K is the message block length and R is the number of redundancy bytes. The redundancy bytes are computed according to Equation 4.8.

$$C(X) = M(X)X^R \bmod G(X) \qquad (4.8)$$

The notation $P(X)$ is used to denote a polynomial P with coefficients in GF(256), representing a string of octets. Regular polynomial arithmetic is applied to these strings, taking into account that the coefficients are added and multiplied according to GF(256) arithmetic, that is, as if they were themselves polynomials with binary coefficients (modulo 2 operations). In Equation 4.8, $M(X)$ is the message polynomial, $C(X)$ is the redundancy polynomial, and $G(X)$ is the generator polynomial (of degree R) of the Reed-Solomon code. This means that $C(X)$ is the remainder after dividing $M(X)$ by $G(X)$. The encoded block consists of the original message octets $M(X)$ followed by the R redundancy octets $C(X)$; this property makes Reed-Solomon a *systematic* code.

In T1.424, both K and R are programmable within certain constraints, but N is always less than or equal to 255. The Reed-Solomon code provides error correction for up to $t = R/2$ *symbol errors*. The number of bit errors in a single errored symbol does not affect the error correcting capability of the code.

At the receiver side, the $2t$ syndromes S_j are computed as follows from the received message $K(X)$.

$$S_j = K(\alpha^j), j = 1...2t \qquad (4.9)$$

By construction, evaluating the received message polynomial at any of the roots of the generator polynomial should yield an all-zero result (syndrome) for any correctly received message. A non-zero syndrome therefore gives us some information about the error polynomial that was "added" (in the GF(256) sense of the word) to the message by the channel. The goal of Reed-Solomon decoding algorithms is to detect and correct these errors.

Decoding the Reed-Solomon codewords consists of finding the error polynomial $e(x) = e_0 + e_1 x + e_2 x^2 + ... + e_{n-1} x^{n-1}$ from the syndrome equations (Equations 4.10) for the received codeword.

$$
\begin{cases}
S_1 = e_{i_1} x_1 + e_{i_2} x_2 + \ldots + e_{i_v} x_v \\
S_2 = e_{i_1} x_1^2 + e_{i_2} x_2^2 + \ldots + e_{i_v} x_v^2 \\
\quad\quad\quad\vdots \\
S_{2t} = e_{i_1} x_1^{2t} + e_{i_2} x_2^{2t} + \ldots + e_{i_v} x_v^{2t}
\end{cases}
\tag{4.10}
$$

Note that in Equation 4.10, both the v error terms $\{e_{i_1},\ldots,e_{i_v}\}$ and the v error locations $\{x_i,\ldots,x_v\}$ are unknown. These equations are non-linear (some of the unknowns have exponents). There are several well-known algorithms, known as Reed-Solomon decoding algorithms, that solve this system of equations and thus correct the received codeword [46].

Linecode Wars and Tone Spacing

"A religious war...".   When there are two nearly equivalent but non-interoperable ways of transmitting a signal, and no single standard yet, you have the ingredients for a linecode war.[7] The pre-2003 situation of VDSL had all that. Linecode wars tend to suck up all the energy and progress of an SDO and take the focus off other technically important matters. Starting a standardization project knowing that there is a lingering linecode issue to be dealt with takes great courage or great foolishness. Successfully completing such a project takes a fair amount of leadership.

The European standardization body ETSI had decided, for its VDSL standard, not to select a single linecode. Its document contains two complete specifications; one for each linecode. Committee T1E1.4, developing an American National Standard, had resigned itself to creating a trial-use standard, valid for two years, with a goal to select a linecode at its expiry. ITU-T Study Group 15 continued to work on the G.vdsl project with the consensus that only one linecode would be specified, but failed to make any significant progress toward selecting one.

The IEEE 802.3ah Task Force was working in the tradition of finding "a single solution for a single problem." The Task Force leadership firmly believed that a single linecode could be selected (after all, the decision threshold was "only 75 percent," as opposed to the [near] unanimity required in other SDOs). Initially, some members must have

---

[7] ...a beast that is believed by many to be second in ferocity to the *connector war* only, which EFM fortunately managed to avoid.

believed that it was a done deal for QAM, as DMT clearly started in an underdog position. The problem faced by the leadership was dealing with the pressure to accept an ETSI-like compromise, and when this pressure seemed to be out of the way, avoiding a linecode debate would take forever, consuming all the project's time in the process.

A two-track approach.   The 10PASS-TS baseline was not approved by the EFM Task Force until a minimal agreement was reached on the way the linecode would eventually be selected. This agreement is captured in the baseline motion (July 2002):[8]

> Adopt presentation *rezvani_1_0302.pdf* (with addition of comments document, *notes_to_editor_1_0302.doc*, with the exception of note 13) as the basis of the first draft. Adopt *omahony_ copper_1_0702.pdf* as the basis for the line code evaluation criteria. The line code selection process recognizes that Committee T1 has a goal of making a VDSL line code decision and will give due weight to that decision.

With this motion, a clear signal was given to T1E1.4 that their work would be taken seriously, but it was also clear that T1E1.4 would have to compress its linecode selection timeline if it wanted to have any say in the 10PASS-TS selection.

In the meantime, the Task Force would progress by developing mutually exclusive subclauses for each of the linecodes in the 10PASS-TS draft, requiring a sadly unavoidable amount of wasted effort. The same approach was adopted for the two competing long-reach solutions, 2BASE-TL and 2PASS-TL.

"Tones, tones, tones...".   In November 2002, support for both 4.3125-kHz and 8.625-kHz tone spacing was made mandatory in the specification for DMT-based 10PASS-TS. This happened in an effort to remove any features with an "optional" character from the draft. Given the fact that 8.625-kHz tone spacing was described as an alternative to the common 4.3125-kHz tone spacing in an *informative* annex of the referenced T1.424/Trial-Use standard, some additional clarification about its status in EFM was requested by a commenter. Surprisingly, the adopted resolution of the comment was not to disallow, but to mandate support for 8.625-kHz tone spacing.

---

[8] The interested reader can find the referenced baseline presentations on the EFM Task Force's Website.

Double tone spacing allows for the use of the same bandwidth with half the size of the IFFT/FFT. This allows a considerable saving in memory and chip area. On the other hand, equalizing becomes more difficult (the tones are twice as wide, so their transfer functions deviate more from the flat assumption), signal processing needs to run twice as fast, and NEXT suppression with respect to 4.3125-kHz systems is lost. With clear and obvious benefits to both having and not having double tone spacing, each company's support for the feature depended mostly on whether they had the capability in silicon. In a commendable concerted effort to close the ranks and focus on the linecode war, DMT supporters left the tone spacing situation unchanged for the remainder of the Task Force review period.

Almost a year later, when the Working Group Ballot on IEEE Draft P802.3ah/D2.0 had started, several comments were submitted asking to remove the mandatory support for 8.625-kHz tone spacing. Although major technical changes at this stage of the standard development process are uncommon, and despite the bias for conservatism imposed by the 75 percent rule, the votes needed to accept these comments were found in the Copper Sub Task Force. Just as surprisingly as it had been introduced, 8.625-kHz tone spacing was removed from the 10PASS-TS draft.

"And the winner is…". At its June 2003 meeting in Anaheim, California, Committee T1E1.4 reviewed the results of the "VDSL Olympics," a set of tests conducted by Telcordia and BTExact, to which four companies had voluntarily submitted VDSL systems or prototypes: Ikanos Communications (DMT), STMicroelectronics (DMT), Infineon Technologies (QAM), and Metalink (QAM). Although only a detailed study of the results can reveal all the strengths and weaknesses of each of the linecodes, it seems fair to say that DMT brought home a better report card. The results were deemed conclusive by the most influential decision makers—the network operators—which led to the approval of a motion selecting DMT as the only linecode for the American National Standard for VDSL and QAM as the only linecode for a new TRQ document. The creation of a TRQ is a technicality that allows a normative text on QAM-VDSL to continue to exist after the expiry of the trial-use standard. The real message of the motion was that as far as T1E1.4 was concerned, DMT had won.

The EFM Task Force met the following week in a specially scheduled two-day interim meeting in Ottawa, Ontario.[9] After one day of

---

[9] Somehow, all the major decisions of the EFM project seem to have been taken in Canada....

resolving comments (against both linecodes!), the second day was reserved for resolving "big ticket items," the biggest of which would be the 10PASS-TS linecode. A clear signal from T1E1.4 was the best anyone could have hoped for (although some participants were obviously unhappy with the outcome), and the question for the Task Force was exactly how much weight was due to the T1E1.4 decision. Quite a bit, apparently, because after two hours of presentations and Q&A, the following motion was finally adopted:

> The 10PASS-TS PHY shall use Multi-Carrier Modulation (MCM), as specified in T1.424/Trial-Use Part 3. In execution of the Editor's Note on page 402 of D1.732, subclauses 62.3 (SCM PMA), 62.5 (SCM PMD), 62.6.4.2 (SCM PMA PICS), 62.6.4.4 (SCM PMD PICS), and subclauses 45.2.1.15 through 45.2.1.28 (SCM registers) shall be removed from the draft. License is granted to the Editors to make necessary changes to text affected by the selection of MCM.

The modified Belgium proposal.   The VDSL linecode selection in T1E1.4 and IEEE 802.3ah led to a feeling that the deadlock in ITU-T Study Group 15 might finally be broken. A majority was indeed found in favor of DMT[10] in a straw poll at the July 2003 Rapporteur's meeting. However, a majority is not enough to make a decision; consensus is required. It wasn't until the January 2004 Rapporteur's meeting in Singapore that a proposal for a way forward with VDSL started to get traction: the Belgium proposal.[11]

The following proposal,[12] a slightly modified version of the original Belgium proposal, received support from Belgium, France, the United Kingdom, and a list of 25 companies at the March 2004 meeting in Millbrae:

> Q4/15 agrees to develop a VDSL Recommendation for consent in April 2004 or sooner with full text specification for DMT in the main body and full text specification for QAM in a normative Annex, with the following in the Scope section: It has been agreed in the ITU-T

---

[10] This is one interpretation of the straw poll results; different questions were asked, and the number of participants in favor of "DMT only" was greater than the number of participants in favor of "QAM only."

[11] Belgium, a 30,520 km^2 kingdom in Western Europe with a population of 10 million, has three cultural communities speaking different languages (Dutch, French, and German). Hence, it is a land of creative compromises. Incidentally, it also has a great reputation for beer and chocolate.

[12] The version shown here has been edited by the author for conciseness.

to develop a subsequent VDSL2 Recommendation that specifies only DMT modulation, and is based on ITU Rec. G.993.1-2004 (VDSL) and ITU Rec. G.992.3 (ADSL2).

Q4/15 agrees to develop a subsequent VDSL2 Recommendation that shall specify only DMT modulation, and shall be based on ITU Rec. G.993.1-2004 (VDSL) and ITU Rec. G.992.3 (ADSL2). Q4/15 also agrees that future VDSL enhancements shall be addressed in this VDSL2 Recommendation.

It is important to understand that an Annex to a Recommendation contains *normative* text. In other words, ITU-T Recommendation G.993.1-2004 attributes the same normative weight to both linecodes; the only difference being that a reader who reads the Recommendation from the front to the back will encounter the DMT specification first. For *future* versions of VDSL, the situation is clear-cut: Only DMT will be used for future enhancements.

A proposal giving less than normative status to QAM would never have obtained consensus in Study Group 15—several countries had already deployed large numbers of QAM-VDSL systems and wanted an ITU-T Recommendation to "protect" this investment, and the suppliers of QAM-VDSL components and systems gladly supported this

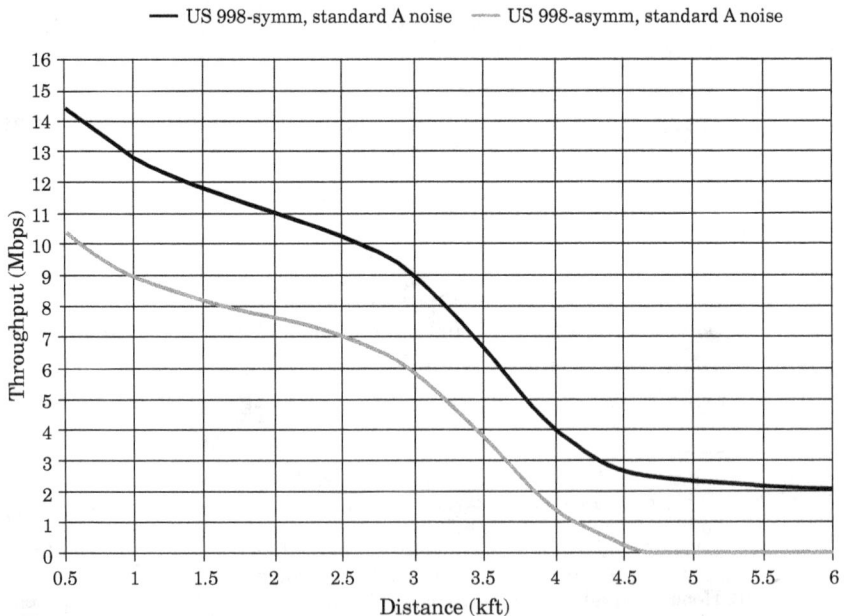

Figure 4.4 Simulated performance of 10PASS-TS (source: EFM Task Force [34]).

point of view. If no compromise had been adopted, the world would have been left with 10PASS-TS as the only VDSL standard in most of the world, which would not have been so bad for the supporters of DMT, but which might also have given the DMT fraction the reputation of "consensus blockers."

Performance

A simulation of the rate-reach characteristics of VDSL, as presented by the supporters of the DMT-VDSL baseline proposal, is shown in Figure 4.4. The Task Force's short-reach objective is the intersection of the 10 Mbps gridline at 2.5 kft gridline[13]—only the upstream bitrate is shown in the plot, because this is the limiting factor for the reach of a symmetric service. The simulation assumes a noise margin of 6 dB and a coding gain of 5.5 dB. Precise performance requirements for 10PASS-TS are provided in Annex 62B of the EFM standard.

10PASS-TS Specifics

Reference Model

The 10PASS-TS PHY specification is built up according to the reference model. PMD and PMA are based on the VDSL PMD and PMS-TC as specified in American National Standard T1.424,[14] with the exceptions and modifications explained in this section.

A very important difference between 10PASS-TS and T1.424 VDSL is the fact that features that are *optional* in T1.424, are *not required for compliance* with 10PASS-TS. This simple statement in the 10PASS-TS standard essentially places all options from T1.424 outside the scope of the IEEE document. This was a deliberate choice to ensure that customers buying equipment labeled "10PASS-TS" would know *exactly* which features they will get. Wherever possible, the initialization mechanisms for T1.424 VDSL options have been left untouched (marked "reserved" in the EFM standard) to allow vendors that do implement them to negotiate their use in an interoperable manner. When any of these T1.424 VDSL options are being used, however, the modems are no longer being used in a 10PASS-TS compliant

---

[13] 750 m is actually 2461 feet. 2500 feet corresponds to the original objective.
[14] This was a draft standard at the time the EFM project started, a trial-use standard by the time the EFM draft went to Working Group Ballot, and will be a standard by the time EFM is published.

mode! Relying on such options in testing of compliance with the 10PASS-TS performance guidelines is therefore not allowed.

10PASS-TS was initially based on the T1.424/Trial-Use standard, which contained both linecodes for VDSL. Committee T1E1.4 and the IEEE 802.3ah Task Force both eventually selected DMT modulation as the only linecode for further standardization. Although this may seem like an obvious choice with hindsight, there was no formal rule or agreement that would have prevented the EFM Task Force from selecting a different linecode if that had been the will of the membership. Although this double defeat was a major setback for companies that had been developing QAM-based VDSL products up to that point, the end of the uncertainty associated with having two linecodes is expected to be an important stimulus for the entire industry.

10PASS-TS Initialization and Management

This section lists all the bits defined in the Standard Information Field for the EFM Copper PHYs. The precise location of each bit in the field is specified in the EFM standard in G.994.1 style (which requires 90 tables). The overview given here is less detailed, but it reveals the specific tree-structure of the transmitted information. Note that the entire 10PASS-TS codepoint tree resides under a dedicated 10PASS-TS SPar(1) codepoint. Unlike American VDSL (American National Standard T1.424), 10PASS-TS is *not* selected as a mode of operation of ITU-T Recommendation G.993.1.

- SPar(1) **10PASS-TS**
  - NPar(2) **Upstream use of 25–138 KHz band** (capability must be present in 10PASS-TS; use is optional)
  - NPar(2) **Downstream use of 25–138 KHz band** (capability must be present in 10PASS-TS; use is optional)
  - NPar(2) **G.997.1—Clear EOC OAM**
  - SPar(2) **Used bands in upstream**
    - NPar(3) **End tone index of band *n*** (12 bits)
    - NPar(3) **Start tone index of band *n*** (12 bits)
  - SPar(2) **Used bands in downstream**
    - NPar(3) **End tone index of band *n*** (12 bits)
    - NPar(3) **Start tone index of band *n*** (12 bits)
  - SPar(2) **IDFT/DFT size**
    - NPar(3) **IDFT/DFT size** (6 bits × 256 points)
  - SPar(2) **Initial length of cyclic extension**

- NPar(3) **Initial sample length of cyclic extension** (10 bits) (*must be 320, 640, or 1280 in 10PASS-TS*)
- SPar(2) **MCM RFI bands**
  - NPar(3) **End tone index of band *n*** (12 bits)
  - NPar(3) **Start tone index of band *n*** (12 bits)

10PASS-TS transceivers use G.994.1 handshaking to exchange band plan information, radio band notches, the size of the IDFT/FFT (number of carriers), and the length of the cyclic prefix. The handshake is followed by a Training state, and a Channel Analysis & Exchange phase, which allows negotiation of additional parameters.[15] The messages in these phases are formatted according to the Special Operation Channel (SOC) protocol and modulated as sets of 4-QAM constellations on all allowed tones.

The Training state is used to negotiate further band plan–specific parameters. This phase consists of the following messages:[16]

O-SIGNATURE
- The bands used in downstream direction
- The bands used in upstream direction
- RFI bands
- Transmit PSD in downstream direction
- Whether UPBO is performed using a maximum PSD or using an upstream PSD mask
- The maximum transmit PSD in upstream direction
- The reference PSD
- The overall length of the window at the transmitter

R-MSG1
- Transmit PSD in upstream
- Echo canceller training flag

O-UPDATEn
- Gain update
- Timing advance correction

---

[15] The parameters negotiated in Training and Channel Analysis & Exchange may override the parameters negotiated in handshake. Both processes are under control of the CO-side.

[16] Messages starting with "O-" are transmitted by the CO-side PHY; messages starting with "R-" are transmitted by the CPE-side PHY. Message descriptor fields are not shown.

O-MSG1

- Final length of the cyclic extension (CE) (*must be 320, 640, or 1280 in 10PASS-TS*)

The Channel Analysis & Exchange phase is used to negotiate the operational parameters for the link. This phase consists of the following messages:

R-MSG2

- Maximal constellation size in upstream (must be 12 in 10PASS-TS)
- Reed-Solomon setting supported by 10PASS-TS-R (see the section "Optional Parameter Values")
- Interleaver setting supported by 10PASS-TS-R (see the section "Optional Parameter Values")
- Detailed interleaver setting description
- Maximal power transmitted
- Maximum interleaver memory
- Maximum number of EOC bytes per frame, upstream (must be 1 in 10PASS-TS, which corresponds to 32 kbps)
- Maximum number of VOC bytes per frame, upstream (must be 1 in 10PASS-TS, which corresponds to 32 kbps)
- Support of express bit swapping
- j_{max} (maximum supported value, must be 0 in 10PASS-TS)

O-MSG2

- Minimal SNR margin
- Maximal constellation size in downstream (must be 12 in 10PASS-TS)
- Reed-Solomon setting supported by 10PASS-TS-O (see the section "Optional Parameter Values)
- Interleaver setting supported by 10PASS-TS-O (see the section "Optional Parameter Values)
- Detailed interleaver setting description
- Maximal power in downstream
- Maximum interleaver delay
- Maximum number of EOC bytes per frame, downstream (must be 1 in 10PASS-TS, which corresponds to 32 kbps)
- Maximum number of VOC bytes per frame, downstream (must be 1 in 10PASS-TS, which corresponds to 32 kbps)
- Support of express bit swapping
- j_{max} (maximum supported value, must be 0 in 10PASS-TS)

R-CONTRACT1
- Proposed downstream contract

O-CONTRACTn
- Downstream contract
- Upstream contract
- EOC capacity
- VOC capacity

R-MARGINn
- SNR margin

O-B&G
- j_{max} (selected value, must be 0 in 10PASS-TS)
- b_i and g_i information

R-B&G
- j_{max} (selected value, must be 0 in 10PASS-TS)
- b_i and g_i information

As indicated above, the parameters that are exchanged during initialization are either forced to a certain value by the EFM standard, or configurable through the Clause 45 management interface (see the section "Management for EFM Copper" in Chapter 3). Certain frequency plan–related parameters can vary per tone, and are hence stored as a large vector in the management information base. These per-tone parameters are particularly useful for the implementation of dynamic spectrum management as a management application; the ability to read the SNR and program the transmit PSD on a 4.3125-kHz granularity allows for substantial optimization of the spectral use of a cable bundle. If desired, frequency plans other than those listed in the section "Frequency Plans" can be specified and activated by means of the per-tone parameters.

To facilitate setting these per-tone parameters, the concept of "tone groups" was introduced in Clause 45: A certain parameter value can be written to any contiguous set of tones with a single write operation. A tone group is set by means of the "10P tone group register" bits (lower tone, upper tone). The actual values of frequency-dependent parameters are written to the "10P tone control parameters register." The values take effect as soon as the corresponding bits in the "10P tone control action register" are triggered.

Reading operations on per-tone parameters can only be performed on individual tones. For this purpose, the read-only "10P tone status register" is used, which returns the value of parameters pertaining to the tone selected as the lower tone in the "10P tone group register." The contents of the "10P tone status register" must explicitly be refreshed through a command in the "10P tone control action register," for the link partner parameters to be updated.

Certain 10PASS-TS parameters have one or more fixed values per the EFM standard, while the initialization messages (handshake and SOC) allow a wider range of values to be negotiated between 10PASS-TS-O and 10PASS-TS-R. This is typically the case for options from American National Standard T1.424 that have been declared outside the scope of EFM. For these parameters, it is up to the 10PASS-TS-O vendor to provide a proprietary interface to these parameters. The 10PASS-TS-R PHY can accept the optional parameter values if capable, or force the link to standard parameters otherwise.

Data-Link Specific Functions

The 10PASS-TS specification has no ambition to be generic at the data link level; it specifies an Ethernet PHY, and nothing more. It is therefore logical that all references to ATM and STM have been removed from the 10PASS-TS specification. This was generally achieved by declaring the corresponding functions and features "out of scope" for EFM and labeling any bits or registers needed for these functions and features "reserved." The advantage of this approach is that it does not explicitly prohibit vendors from including this functionality in products and making multi-mode devices.

Conversely, a number of new registers had to be defined for the EFM-specific TPS-TC functions (these can be found as the TC-specific and PCS-specific registers in Clause 45 of EFM). An EFM-TC-specific mechanism was selected to allow the TPS-TCs on both sides of the link to communicate loss-of-sync to each other. This is done by means of a special idle character in the 64/65-octet encapsulation. The (at least for 10PASS-TS) more logical choice of using indicator bits for this purpose was reversed because it could not be supported by existing ESHDSL silicon on which the first 2BASE-TL implementations were to be based.

Optional Parameter Values

As noted before, features that are optional in American National Standard T1.424 are not required for compliance with 10PASS-TS. This has the following consequence for 10PASS-TS:

- No support for the fast path, hence no support for dual latency.
- Reed-Solomon parameters are limited to mandatory values $(N, K) = (144, 128)$ and $(N, K) = (240, 224)$.
- Interleaver parameters are limited to mandatory values $I = 36$, $M \in [2, 52]$ for $(N, K) = (144, 128)$ and $I = 36$ and $M \in [2, 52]$ for $(N, K) = (240, 224)$. (Additionally, the interleaver can be turned off, which makes the slow path behave like a fast path. This is implicit in the description of the interleaver in American National Standard T1.424.)
- Only 1 VOC byte and 1 EOC byte per frame are supported.
- No interpolation is applied to bitloading (b_i) and gain scaling (g_i) values (i.e., j_{max} is fixed to 0).

Additionally, the number of carriers is fixed at 4096 (this corresponds to the maximum in T1.424), and the number of bits per carrier supported in upstream and downstream is fixed at 12 (in T1.424, this number can be anywhere between 8 and 15). The choices for the length of the cyclic extensions are limited to $m = 10$, $m = 20$, and $m = 40$. Other values for these parameters are declared out of scope.

Dual Latency and Framing

The basic characteristics and benefits of dual latency are explained in Chapter 3. Support for dual latency is optional in T1.424. To preserve the architectural simplicity of Ethernet, and to limit the number of optional features to the strict minimum, dual latency was not kept in the EFM standard.

As a result, the PMA frame for 10PASS-TS is identical to the PMS-TC frame for VDSL with only the slow buffer present. Note that Network Timing Reference (NTR) is not used in 10PASS-TS and that new indicator bits have been defined specifically for EFM.

The requirement on the VDSL PMS-TC to offer 64 kbps bitrate granularity at the $\alpha(\beta)$-interface, as imposed by the framing algorithm, must be reconciled with the implicit requirement to offer 64-kbps bitrate granularity at the level of the MII, imposed by the payload data rate registers in management. Taking the 64/65-octet encapsulation into account, it is clear that a given bitrate at the MII can only be achieved by setting the rate at the $\alpha(\beta)$-interface at least 1.56 percent (with a minimum of 64 kbps) higher than the one required at the MII, and using the MAC-PHY rate matching function for further fine-tuning.

Profiles and Internationalization

The 10PASS-TS standard specifies profiles[17] for a large number of "typical" deployment scenarios, including all known settings that are mandatory under various local regulations. To fully specify the behavior of a 10PASS-TS modem, a profile must be selected for the band plan and PSD, one for the reference PSD for UPBO, one for band notches, and one for payload data rates.

Band plan and PSD. Adherence to a particular band plan and PSD mask is mandatory when the service is deployed in a public network. In a private network, the network operator may select a band plan that is most suitable to offer the desired data rates on the existing network infrastructure.

Reference PSD. The reference PSD for upstream power back-off should be selected on the basis of the band plan used in the cable plant and the topology of the deployment (cabinet or CO).

Band notches. Notches reduce the output power in certain narrow frequency bands, to avoid electromagnetic egress that might interfere with radio transmissions. Activation of certain region-dependent notches may be mandatory when the transmission cable is not buried underground.

Payload data rates. The data rate profiles provided to the end user range between 2.5 Mbps and 50 Mbps (upstream) or 100 Mbps (downstream).

This wide range of different possible configurations may be a challenge for implementers, but it makes 10PASS-TS the first truly international VDSL standard; a 10PASS-TS–compliant system can indeed be deployed in the regions previously covered separately by ANSI and ETSI standards. The ITU-T recommendation for VDSL does not spec-

---

[17] See "Management for EFM Copper" in Chapter 3 for a discussion of EFM Copper management and profiles.

ify an internationally usable system, as it will only require the implementation of one of the regional annexes for compliance.

This international nature of the 10PASS-TS standard is at once a strength and a handicap. Although the standard specifies the 10PASS-TS transceiver completely, certain parameters of the electrical interface (i.e., termination impedance,[18] output signal balance requirement, connector specifications) may actually vary between deployment regions due to local regulations. Covering all possible cases in the standard was deemed unpractical; implementing a system that is deployable everywhere may be just as unpractical.

On an EFM Copper link, the CO-side system is the one that selects the profiles; it is therefore possible to build a CO-side system that supports only a limited set of profiles (e.g., to serve a particular geographical region), which cannot be distinguished from a fully compliant 10PASS-TS-O by a compliant 10PASS-TS-R. As CO equipment is typically sold for use in a specific country or region, and not moved around much after it is put in service, it is to be expected that CO-systems with "regionally limited compliance" will be brought to the market.

For the CPE-side device, supporting all profiles is more important; it must be capable of supporting whichever standard profile the CO-side system dictates, regardless of whether the customer bought the modem in a duty-free shop at Hong Kong airport or at an electronics store in the United States.

TABLE 4.1 Default Profile for 10PASS-TS

Parameter	Value
Downstream Bitrate	10 Mbps
Upstream Bitrate	10 Mbps
Frequency Plan	Band Plan A (as in T1.424)
PSD	Cabinet mask M1
Reed-Solomon configuration	(240, 224)
Interleaver configuration	$I = 30, M = 62$

---

[18] In VDSL, the termination impedance is region dependent, with a value of 135 Ω to be used in Europe and a value of 100 Ω to be used in North America. VDSL2 will specify a termination impedance of 100 Ω everywhere, like ADSL does, which is not expected to cause any noticeable service degradation.

The support of profiles is further regulated by the "performance guidelines" in the EFM standard. For a set of 22 complete profiles (i.e., combinations of a data rate pair, a frequency plan, a reference PSD, and RFI notches) and specific crosstalk environments, the standard specifies the minimum reach that must be met on a test cable. To comply with the standard, a 10PASS-TS PHY must pass all test cases, without use of any options outside the scope of the standard. This will of course be impossible for certain 10PASS-TS based systems that are only regionally compliant, as described. The numbers of the test cases listed in the standard can be used to select a complete profile via the MIB objects defined in Clause 30 and the corresponding IETF MIB objects.

It may be noted that the VDSL2 recommendation will also use profiles, but in a slightly different manner. Instead of assuming, as 10PASS-TS does, that every transceiver has exactly the same capabilities from which different operational parameters can be selected, VDSL2 will use "profiles" to designate the different sets of capabilities that the VDSL2 transceiver may have. In that way, the VDSL2 specification will actually specify multiple classes of transceivers, or—in Ethernet language—multiple different "port types."

The Long-Reach Solution: 2BASE-TL

"Like it or not, the business environment is where long reach EFM is needed now."
—THE 2BASE-TL BASELINE PROPOSAL [26]

Introduction to SHDSL

Synchronous Digital Hierarchy

The Plesiochronous Digital Hierarchy (PDH) and Synchronous Digital Hierarchy (SDH/SONET) specify different levels of multiplexing for data transport. T1 and E1 links are first-order multiplexes of digitized telephone lines, used for trunking between a private exchange and a central office (CO). Each digitized telephone line has an uncompressed bit rate of 64 kbps, corresponding to 8000 audio samples per second, with a size of 8 bits each.[1]

The American variant T1 has a rate of 24×64 kbps = 1536 kbps. The European variant E1 has a rate of 32×64 kbps = 2048 kbps. The original implementations of T1/E1 links used a robust but spectrally inefficient modulation scheme, generating particularly nasty crosstalk.

---

[1] For comparison, an audio CD uses a sampling rate of 44,100 samples per second with a size of 16 bits each, which explains the difference in sound quality between a telephone and a CD. The sampling rate is twice the highest frequency that can be reproduced (4 kHz for a phone, 22.5 kHz for a CD; the latter spanning the entire frequency range audible to the human ear). The size of the samples determines the signal-to-noise ratio of the reproduced sound.

High-Speed Digital Subscriber Line (HDSL), Synchronous Digital Subscriber Line (SDSL), and their successors Single-Pair High-Speed Digital Subscriber Line (SHDSL) and HDSL2 were introduced to offer T1 and E1 services more efficiently and with less crosstalk.

Pulse Amplitude Modulation

Pulse Amplitude Modulation (PAM) codes a set of bits into the amplitude of a pulse. In SHDSL, 16-level PAM is used.[2] Before being mapped onto PAM symbols, the serial bit stream of user data is first scrambled and then sent through a convolutional encoder. The convolutional encoder transforms m input bits into $m+1$ output bits. This implies that the 4-bit PAM symbol carries the equivalent of 3 user data bits.

SHDSL supports different bitrates r [kbps], as defined by Equation 5.1, by varying the symbol rate (baud rate), and thus the spectral width of the transmit signal, symbolized in the following equation by parameters n ($3 \le n \le 36$) and i ($0 \le i \le 7$).

$$r = n \times 64 + i \times 8 \tag{5.1}$$

The symbol rate f_{sym} [Hz], required to support a given bitrate (data rate r + 8 kbps framing overhead), is determined by the number of uncoded bits per symbol (3 bits/symbol for SHDSL).

$$f_{sym} = 1000 \times (r + 8)/3 \tag{5.2}$$

Without pulse shaping, a rectangular PAM pulse has a sync-shaped[3] power spectral density (PSD), with a highly energetic main lobe between DC and $1/f_{sym}$, and slowly decreasing sidelobes. Pulse shaping results in a steeper roll-off of the PSD, which is necessary to comply with the approved PSD masks, which ensure spectral compatibility with other basis systems.

The achievable bit rate is limited by the channel transfer function and the amount of crosstalk present (mostly self-NEXT); the longer the loop over which SHDSL operates, the narrower the usable frequency band, and the lower the achievable bit rate becomes. The limitation of

---

[2] Extended to 32-PAM in the next-generation "enhanced SHDSL," allowing for 4 bits per symbol (ITU-T Recommendation G.991.2, amended).

[3] Proportional to $\dfrac{\sin^2 x}{x^2}$.

useful bandwidth with increasing loop lengths automatically ensures spectrum compatibility of SHDSL with the basis systems (see the section "Spectrum Management" in Chapter 3).

The bitrates defined by Equation 5.1 are all multiples of 8 kbps, with multiples of 64 kbps as a special case for $i=0$. This fact reveals SHDSL's origin as a replacement for HDSL, which is used to offer T1/E1 trunk line services.

At the PMS-TC level, user data is framed into data frames,[4] which have a nominal duration of 6 ms. Each of these data frames contains 4 payload blocks, each of which is further divided into 12 sub-blocks. A simple calculation shows that there are 8000 sub-blocks in a second; this means that a sub-block is the ideal container to transport the samples for one or more 64 kbps data streams.

The SHDSL frame furthermore carries indicator bits and an embedded operations channel (EOC) for management purposes,[5] stuff bits to maintain byte alignment if $i \neq 0$, and a CRC.

The PSD masks for symmetric SHDSL operation (the only type of operation used by 2BASE-TL) are given by the following equation, for $f < f_{int}$.

$$\text{PSDMASK}_{\text{SHDSL}}(f) = 10^{\frac{-PBO}{10}} \times \frac{K_{\text{SHDSL}}}{135} \times \frac{1}{f_{\text{sym}}} \times \frac{\left[\sin\left(\frac{\pi \cdot f}{Nf_{\text{sym}}}\right)\right]^2}{\left(\frac{\pi \cdot f}{Nf_{\text{sym}}}\right)^2} \times \frac{1}{\left(1+\frac{f}{f_{\text{3dB}}}\right)^{2 \times Order}} \times 10^{\frac{\text{MaskOffsetdB}(f)}{10}}$$

$$(5.3)[6]$$

For $f_{int} < f < 1.1$ MHz, $\text{PSDMASK}_{\text{SHDSL}}(f) = 0.5683 \ 10^{-4} \ f^{-1.5}$. f_{int} is defined as the frequency where the two functions defining PSD-MASK$_{\text{SHDSL}}(f)$ intersect. MaskOffsetdB(f) is defined by Equation 5.3. These definitions lead to the PSD masks shown in Figure 5.1.

$$\text{MaskOffsetdB}(f) = \begin{cases} 1+0.4\times\dfrac{f_{\text{3dB}}-f}{f_{\text{3dB}}} & f < f_{\text{3dB}} \\ 1 & f \geq f_{\text{3dB}} \end{cases} \qquad (5.4) \ [7]$$

---

[4] These have nothing to do with "data frames" defined at the TPS-TC level.
[5] These overhead channels must not be confused with the EFM OAM messages described in Chapter 8.
[6] Reproduced with the kind permission of ITU.
[7] Reproduced with the kind permission of ITU.

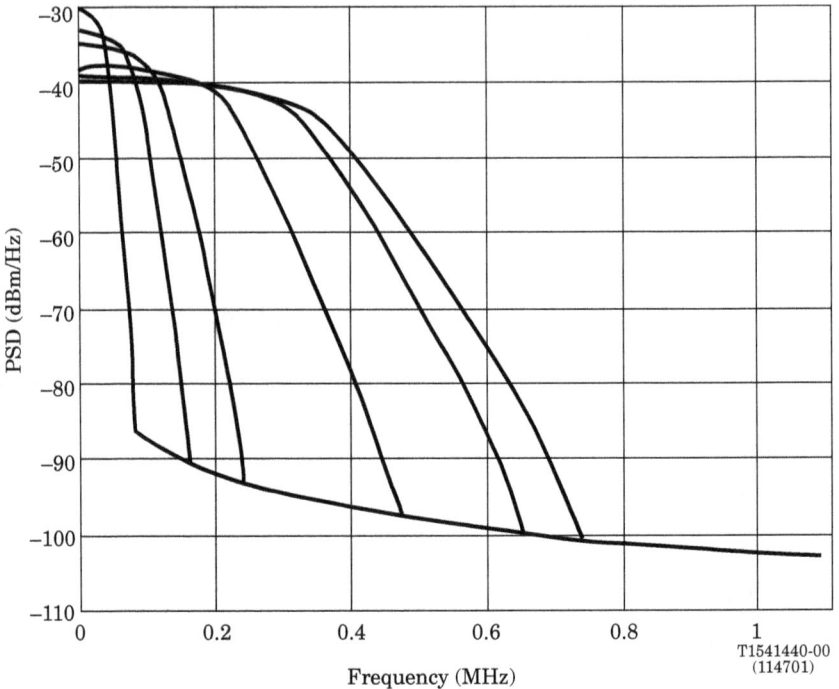

Figure 5.1 ITU-T recommendation G.991.2 PSD Masks for symmetric services, for use in Region A.[8]

Spreading the available transmit power over more bandwidth when it is available (i.e., increasing the symbol rate), is a beautiful principle: As we recall from the Shannon-Hartley theorem (Equation 3.4), the capacity of the link increases linearly with the available bandwidth and only logarithmically with the signal-to-noise ratio (SNR).

2BASE-TL Specifics

Reference Model

The 2BASE-TL Physical Layer entity sublayer (PHY) specification is built up according to the reference model shown in Figure 3.3. Physical Medium Dependent sublayer (PMD) and Physical Medium Attachment sublayer (PMA) are based on the original SHDSL PMD and PMS-TC, as specified in ITU-T Recommendation G.991.2,[9] with the exceptions and modifications explained in this section.

---

[8] Reproduced with the kind permission of ITU.
[9] The edition referenced by IEEE Std 802.3ah-2004 is G.991.2 Edition 2001, as amended by Amendment 1 (2001).

Enhanced SHDSL

During the course of the development of the Ethernet in the First Mile (EFM) standard, a number of enhancements to the existing SHDSL standards were approved in T1E1.4 and ITU-T Study Group 15. These enhancements, which support higher bitrates, were imported into 2BASE-TL.

First, enhanced SHDSL and 2BASE-TL supports new PSDs with a greater bandwidth, accommodating higher symbol rates ($3 \le n \le 89$).

Second, enhanced SHDSL and 2BASE-TL support 32-level TC-PAM (4 data bits encoded in 5 bits) in addition to the 16-level TC-PAM of traditional SHDSL, which allows a 33 percent increase in bitrate on loops with a sufficiently high SNR. Alternatively, it allows the operator to use a lower symbol rate for a given bitrate, as the denominator in Equation 5.2 increases, which improves spectral compatibility. This enhancement is mainly useful on short loops; however, on such loops, 10PASS-TS should be considered as a more flexible alternative.

A simulation of the rate-reach characteristics of enhanced SHDSL or 2BASE-TL is shown in Figure 5.2. The horizontal dashed line represents the transition from traditional SHDSL to enhanced SHDSL. The figure represents performance on a 0.5mm pair with only additive white gaussian noise (AWGN). The simulation assumes a noise margin of 5 dB and a coding gain of 5 dB. Note also that no T1 disturbers were included in the simulation; these are known to have a severe

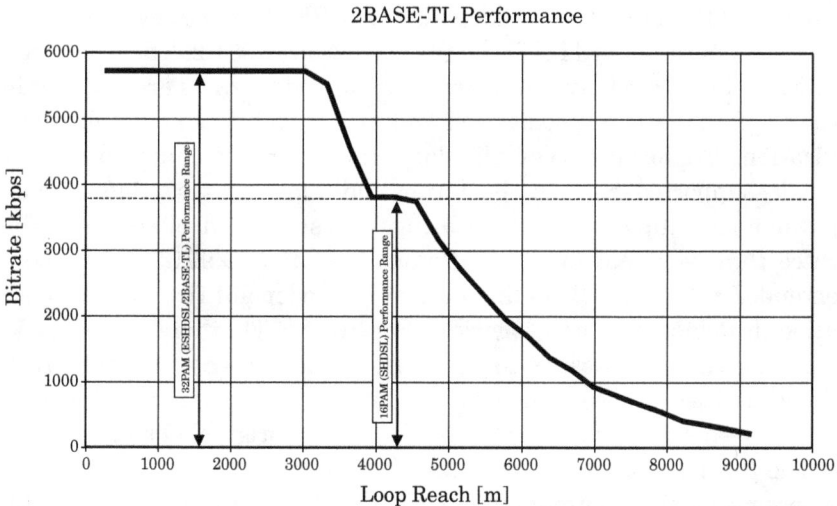

Figure 5.2   Simulated performance of 2BASE-TL.

detrimental effect on the performance of (enhanced) SHDSL links. Precise performance requirements for 2BASE-TL are provided in Annex 63B of the EFM standard.

2BASE-TL Initialization and Management

2BASE-TL transceivers may optionally go through a phase called Preactivation, before starting the Core Activation. Preactivation consists of a special G.994.1 handshaking session initiating the Power Measurement Modulation Session (PMMS, a.k.a. line probing), in which the capacity of the line is investigated through an uncoded 2-PAM test signal representing a scrambled sequence of all ones. The test signal is transmitted on request of the receiver and may include sequences at different symbol rates.

After the PMMS, a second G.994.1 handshaking session is used to determine the operational parameters that shall be used during the Core Activation (training) and Data Mode. Training consists of an uncoded 2-PAM signal, at the symbol rate selected for data mode operation, which is automatically followed by data mode.

This section lists all the bits defined in the Standard Information Field for 2BASE-TL. The precise location of each bit in the field is specified in the EFM standard in G.994.1 style. The overview given here is less detailed, but it reveals the specific tree-structure of the transmitted information.

Note that the entire 2BASE-TL codepoint tree resides under a dedicated 2BASE-TL SPar(1) codepoint—2BASE-TL is *not* selected as a mode of operation of ITU-T Recommendation G.991.2. There is a significant difference between the way in which data rates are encoded in the original G.shdsl specification and in the "Enhanced SHDSL" specification. The former uses individual NPar(3) bits to enumerate every single supported bitrate; the latter additionally uses min/step/max parameter 3-tuples to indicate ranges of supported bitrates for the rates that were not in ITU-T Recommendation G.991.2-2001; these extended rates are only defined under the codepoint that selects operation in Region 1 (North America). For 2BASE-TL, there are no backward-compatibility issues to deal with, which is why the EFM Task Force decided to use min/step/max bitrate coding for *all* supported bitrates—not just the ones that are above the original G.shdsl range.

The 3-tuple represents all rates of the form $64 \times (r_{min} + ks)$ [kbps] where r_{min} is the minimum value, s is the step value, and k is an integer such that $0 \le k \le (r_{max} - r_{min}) / s$. These 3-tuples are programmed via the "2B PMD parameters registers" specified in Clause 45 (see also "Managment for EFM Copper" in Chapter 3). If line probing is enabled, the result of the probing further limits the data rate ranges that can be used by the 2BASE-TL PHYs: Of the rates specified by management, only those that pass the line probing are indicated in handshake messages.

- SPar(1) **2BASE-TL**
- NPar(2) **2BASE-TL training mode**
- NPar(2) **2BASE-TL PMMS mode**
- NPar(2) **2BASE-TL Band A operation**
- NPar(2) **2BASE-TL Band B operation**
- NPar(2) **Regenerator silent period**
- NPar(2) **SRU** (for use with regenerators, outside the scope of EFM)
- NPar(2) **Diagnostic Mode** (for use with regenerators, outside the scope of EFM)
- SPar(2) **2BASE-TL Downstream training parameters**
  - NPar(3) **Downstream PBO** (dB) (5 bits × 1.0 dB)
- SPar(2) **2BASE-TL Downstream training parameters 16-TCPAM**
  - NPar(3) **Downstream base data rate 16-TCPAM minimum** (7 bits)
  - NPar(3) *Downstream base data rate 16-TCPAM maximum* (7 bits)
  - NPar(3) **Downstream base data rate 16-TCPAM step** (7 bits)
  Multiple instances of the parameter 3-tuple may be present.
- SPar(2) **2BASE-TL Downstream training parameters 32-TCPAM**
  - NPar(3) **Downstream base data rate 32-TCPAM minimum** (7 bits)
  - NPar(3) **Downstream base data rate 32-TCPAM maximum** (7 bits)
  - NPar(3) **Downstream base data rate 32-TCPAM step** (7 bits)
  Multiple instances of the parameter 3-tuple may be present.
- SPar(2) **2BASE-TL Upstream training parameters**
  - NPar(3) **Upstream PBO** (dB) (bits 5-1 × 1.0 dB)
- SPar(2) **2BASE-TL Upstream training parameters 16-TCPAM**
  - NPar(3) **Upstream base data rate 16-TCPAM minimum** (7 bits)
  - NPar(3) **Upstream base data rate 16-TCPAM maximum** (7 bits)
  - NPar(3) **Upstream base data rate 16-TCPAM step** (7 bits)
  Multiple instances of the parameter 3-tuple may be present.
- SPar(2) **2BASE-TL Upstream training parameters 32-TCPAM**

- NPar(3) **Upstream base data rate 32-TCPAM minimum** (7 bits)
- NPar(3) **Upstream base data rate 32-TCPAM maximum** (7 bits)
- NPar(3) **Upstream base data rate 32-TCPAM step** (7 bits)
Multiple instances of the parameter 3-tuple may be present.
- SPar(2) **2BASE-TL Downstream PMMS parameters**
  - NPar(3) **Downstream PBO** (dB) (5 bits × 1.0 dB)
  - NPar(3) **Downstream PMMS duration** (6 bits × 50 ms) or unspecified
  - NPar(3) **Downstream PMMS scrambler polynomial**
  - NPar(3) **Worst-case PMMS target margin** (6 bits × 1.0 dB – 10 dB) or unspecified
  - NPar(3) **Current-condition PMMS target margin** (6 bits × 1.0 dB – 10 dB) or unspecified
  - NPar(3) **Transmit silence**
- SPar(2) **2BASE-TL Downstream PMMS rates**
  - NPar(3) **Downstream base data rate 32 TC-PAM minimum** (7 bits)
  - NPar(3) **Downstream base data rate 32 TC-PAM maximum** (7 bits)
  - NPar(3) **Downstream base data rate 32 TC-PAM step** (7 bits)
Multiple instances of the parameter 3-tuple may be present.
- SPar(2) **2BASE-TL Upstream PMMS parameters**
  - NPar(3) **Upstream PBO** (dB) (5 bits × 1.0 dB)
  - NPar(3) **Upstream PMMS duration** (6 bits × 50 ms) or unspecified
  - NPar(3) **Upstream PMMS scrambler polynomial**
  - NPar(3) **Worst-case PMMS target margin** (6 bits × 1.0 dB – 10 dB) or unspecified
  - NPar(3) **Current-condition PMMS target margin** (6 bits × 1.0 dB – 10 dB) or unspecified
  - NPar(3) **Transmit silence**
- SPar(2) **2BASE-TL Upstream PMMS rates**
  - NPar(3) **Upstream base data rate 32 TC-PAM minimum** (7 bits)
  - NPar(3) **Upstream base data rate 32 TC-PAM maximum** (7 bits)
  - NPar(3) **Upstream base data rate 32 TC-PAM step** (7 bits)
Multiple instances of the parameter 3-tuple may be present.
- SPar(2) **2BASE-TL Downstream framing parameters**
  - NPar(3) **Sync word** (14 bits)
  - NPar(3) **Stuff bits** (2 bits)
- SPar(2) **2BASE-TL Upstream framing parameters**
  - NPar(3) **Sync word** (14 bits)
  - NPar(3) **Stuff bits** (2 bits)

Byte-Oriented Mode

As was indicated when we introduced Equation 5.1 for the bitrates supported by the SHDSL PHY, the parameter i allows fine-tuning of the bitrate in 8 kbps increments. This corresponds to increments of 48 bits per frame, or 1 bit per sub-frame (there are 4 payload blocks with 12 subframes each in a frame). In other words, the parameter i represents single-bit granularity for the capacity of a subframe.

2BASE-TL is an Ethernet PHY, which is used to transport 802.3 MAC frames only. Such frames always come in integer numbers of bytes, and 64/65-octet encapsulation maintains the byte synchronicity. Therefore, 2BASE-TL does not require bit-level granularity for the PMA subframe size, and setting $i = 0$ was made mandatory in EFM.

Note that the TC sublayer used in EFM Copper relies on preservation of the byte boundaries in the data, which implies that a SHDSL PMD that does not preserve byte boundaries across the link cannot be used to build a compliant 2BASE-TL system.

Four-Wire Mode

ITU-T Recommendation G.991.2 specifies an optional mode of operation over four wires (i.e., two pairs), which offers twice the capacity of single pair. In EFM, this mode is obsoleted by PME aggregation, which allows the combination of up to 32 2BASE-TL links into a single logical Ethernet pipe.

Plesiochronous Mode

ITU-T Recommendation G.991.2 supports four clock synchronization modes:

Plesiochronous. CO uses local oscillator, customer premises equipment (CPE) uses received symbol clock.

Plesiochronous with timing reference. CO uses network reference clock; CPE uses received symbol clock.

Synchronous. CO uses transmit data clock or network reference clock; CPE uses received symbol clock.

Hybrid. CO uses transmit data clock; CPE uses received symbol clock.

Out of these modes, 2BASE-TL only supports synchronous mode, using the transmit data clock as a reference.

Signal Regenerators

Signal regenerators are devices that terminate an SHDSL link on a first wire segment, and re-encode the same data as SHDSL symbols onto a second wire segment. Such devices could be thought of as two-port repeaters, and they are usually employed to span a greater distance than would be possible with a simple CO/CPE pair. The presence of signal regenerators between the CO and the end user's CPE has to be detected during initialization, and special handshake codepoints have been defined to ensure adequate configuration of each of the SHDSL link segments.

Adding support for signal regenerators to the 2BASE-TL specification would have placed a considerable burden on the architectural simplicity of the standard. The Ethernet standard already specifies repeaters for 10 Mbps, 100 Mbps, and 1000 Mbps networks (better known as "hubs"), but these are multiple-port half-duplex devices,[10] which bear little resemblance to the SHDSL signal regenerator. Specifying a new repeater-like device for 2BASE-TL was generally considered undesirable.

On the other hand, completely removing support for signal regenerators in 2BASE-TL would clip 2BASE-TL's wings. Without signal regenerators, the reach of 2BASE-TL would be limited to the length of a single link segment, which is by definition much less than the reach of SHDSL with signal regenerators.

To resolve the dilemma, the Task Force decided to place the use of G.991.2 signal regenerators "outside the scope" of the EFM standard. Additionally, a number of codepoints were added to the 2BASE-TL handshaking tree to negotiate the use of signal regenerators. The rationale of this approach is to allow 2BASE-TL transceivers to use existing G.991.2 signal regenerators, without having to specify the details of a new 2BASE-TL signal regenerator.

Profiles and Internationalization

The ITU-T recommendation for SHDSL follows the ITU-T tradition to separate out regional requirements in different normative Annexes, only one of which must be complied to at a time. In this way, a single recommendation really specifies different modems for different regions.

---

[10] Their use is increasingly being abandoned in favor of low-cost Ethernet switches.

From the onset of the EFM project, it was the goal to create a "single solution for a single problem," which made the use of different mutually exclusive regional Annexes impossible. 2BASE-TL was therefore specified as a single type of modem, which has to be capable of complying with two sets of regional requirements (corresponding to region A— North America, and region B—Europe of ITU-T Recommendation G.991.2).

The compliance region can be selected at the CO-side by setting the "2B general parameter register" through the MDIO interface (if present). Alternatively, a management agent can be used to select one out of ten completely defined profiles (see the section "Management for EFM Copper" in Chapter 3), half of which comply with Region A requirements and the other half with Region B requirements. The profiles further differ in the data rates they offer, ranging from 512 kbps to 5696 kbps. Whichever mechanism is used to select the compliance region at the CO-side, the choice is communicated to (and imposed on) the CPE during handshake.

TABLE 5.1 Default Profile for 2BASE-TL

Parameter	Value
Data Rate	5696 kbps
Line Rate	5704 kbps
Transmit Power	14.5 dBm
PSD	Region B
Constellation	32-TCPAM

The Defeated Alternative: 2PASS-TL

"2PASS-TL" is not a part of the EFM standard. It is the name of the ADSL-based long-reach proposal that was defeated by 2BASE-TL in the long-reach linecode war (see the section "Long-Reach Objective" in Chapter 3). There are several presentations in the EFM archives on the web that describe 2PASS-TL in great detail, and these were transformed into text, which was actually part of the EFM specification up to Draft 1.1a. At the risk of being accused of rewriting history, I would dare to claim that 2PASS-TL isn't really gone. There are two reasons to believe that services very similar to 2PASS-TL will indeed become available, despite the lack of a 2PASS-TL standard.

First, the way the EFM Copper TC sublayer and PCS are specified in IEEE 802.3ah, it is an open invitation for any xDSL chipset vendor to add the EFM Copper TC sublayer and PCS to the xDSL lower sublayers of any flavor. Although the installed base of ADSL modems is nearly 100 percent ATM-based, it would be surprising if EFM-style "Ethernet-over-ADSL" modems would not appear in the future. One of the reasons for adding preemption to the EFM-TC variant used in ITU-T, is that it is needed to preserve the quality of a packetized-voice service over a very low-bitrate link, such as the upstream of an ADSL link.

Second, the 10PASS-TS specification provides so much flexibility in the allocation of the transmit spectrum that all of the "ADSL spectra" (as shown in Figure 5.3) can be applied to 10PASS-TS. The EFM standard includes an Annex dedicated to examples of usage of 10PASS-TS, which shows the settings for "Plan A with variable low-frequency region," a frequency plan that can be made to look like the ADSL band plans from 25 kHz up. Note, however, that if 10PASS-TS is used at typical ADSL reaches (1.5–5 km), the benefit of the orthogonality between upstream and downstream carriers is lost (see below).

Introduction to ADSL

Overview. Asymmetric Digital Subscriber Line (ADSL) is the archetypical *residential* xDSL flavor. Originally designed for video-on-demand, it supports a high downstream capacity (up to 8 Mbps) combined with a limited upstream capacity (up to 1 Mbps). This capacity setup proved to be ideal for late-1990s/early-2000s Internet usage, where subscribers would typically read web pages (http requests generate limited upstream traffic, the actual content generates the high downstream traffic) or download large files. As a service, it could be marketed as an "upgrade" from dial-up Internet service, with the increased speed and the permanent availability of the telephone (POTS or ISDN) service on the same line (by means of spectral filters or "splitters") as key differentiators. Current threats to the continued success of ADSL are the increased success of peer-to-peer traffic—symmetric by nature—and the fact that residential video-on-demand is catching on very slowly and may not break through before ADSL is obsoleted by other technologies.

ADSL has become successful within the paradigm of ATM-based access networks, where data pipes with well-controlled capacity and Quality-of-Service (QoS) parameters connect each subscriber to the service provider's network. The majority of deployed ADSL, therefore,

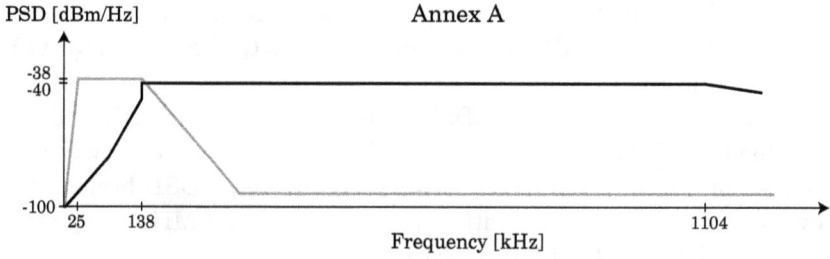

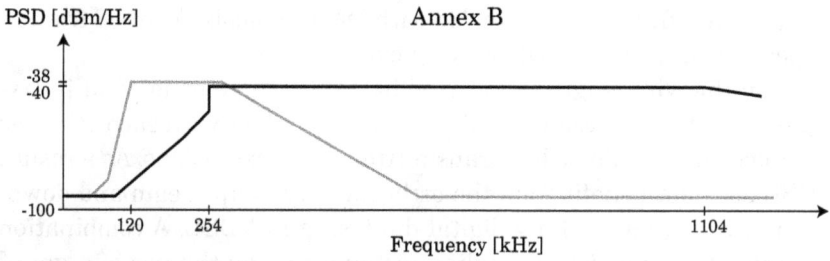

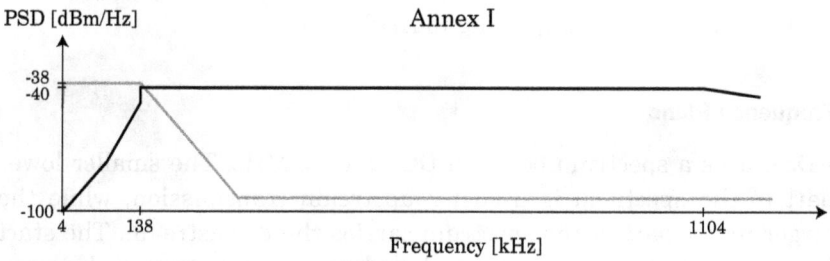

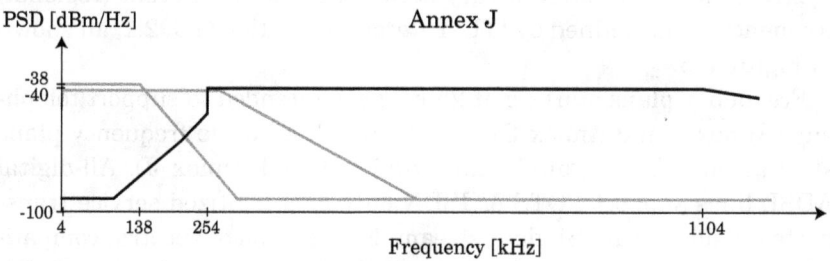

Figure 5.3 ADSL frequency plans, as defined in ITU-T Recommendations G.992.1 and G.992.3.

has an ATM-TC as the transport protocol-specific sublayer. At the relatively low speeds of ADSL links, the hooks for QoS provided by ATM are no luxury [6].

The modulation used for ADSL is Discrete Multi-Tone (DMT), as described in Chapter 4. about 10PASS-TS.[11] There are, however, a number of subtle differences between ADSL and VDSL besides the obvious difference in bandwidth (ADSL uses up to 1.1 MHz, as opposed to the 12-MHz bandwidth of VDSL).

In the PMS-TC (equivalent to the PMA in Ethernet speak) data are mapped onto frames that coincide with DMT symbols. After 68 frames, a special frame is inserted carrying extra overhead.

Given the wire lengths supported by ADSL, there is no point in trying to align the upstream and downstream symbols with each other, as too much time would be lost transmitting cyclic extensions. As a result, ADSL does not benefit from the orthogonality of upstream and downstream, which allows fully digital duplexing in VDSL. A combination of analog filtering, echo canceling techniques, and the use of a guard band (a set of tones around the split frequency that are not used) must be applied to achieve duplexing in ADSL.

Frequency Plans

ADSL uses a spectrum between DC and 1.1 MHz. The smaller lower part of the spectrum is used for upstream transmission, while the larger upper part of the spectrum carries the downstream. The start frequency of the upstream band and the split frequency between upstream and downstream vary between the four different (regional) frequency plans defined by ITU-T Recommendation G.992.1, all shown in Figure 5.3.

Frequency plans starting at 25 kHz are intended to support telephony (Annex A and Annex B) on the same line, while frequency plans starting at DC are "all-digital" (Annex I and Annex J). All-digital ADSL lines support POTS or ISDN only as a digitized service transported inside the ADSL data stream. For reasons of spectral compatibility, the split frequency is at 138 kHz for regions where no ISDN exists in the bundle (Annex A and Annex I), or at 276 kHz for regions where ISDN does exist (Annex B and Annex J). These frequency plans make ADSL suited for subscriber loops of up to around 5 km.

---

[11] ADSL standardization has gone through a DMT versus CAP/QAM linecode war of its own, but this was long before the concept of "Ethernet in the First Mile" was ever brought up.

Beyond ADSL. ADSL comes in different sub-flavors. In ITU-T, there are currently five recommendations for ADSL transceivers:

G.992.1. Full-Rate ADSL, the "classic" flavor, as developed under the G.dmt project.

G.992.2. Splitterless ADSL, using only half the number of carriers and requiring no splitters to allow POTS on the same line; also known under its project name "G.lite."

G.992.3. ADSL2, the enhanced classic, as developed under the G.dmt.bis project. It now includes an annex specifying Reach-Extended ADSL (READSL).

G.992.4. G.lite.bis, which is G.lite with the same enhancements as ADSL2.

G.992.5. ADSL2plus, a version of ADSL2, which supports twice the number of carriers; this doubles the aggregate bitrate on short loops.

Besides these, several regional ADSL standards exist. The original American National Standard for ADSL, T1.413, has been obsoleted by new American National Standard, which points to ITU-T Recommendation G.992.1.

2PASS-TL Specifics

Reference model. The proposed 2PASS-TL PHY specification was built up according to the reference model. PMD and PMA are based on the ADSL2 PMD and PMS-TC, as specified in ITU-T Recommendation G.992.3,[12] with the exceptions and modifications explained in this section.

2PASS-TL includes the same architectural changes with respect to ADSL2 as 10PASS-TS and 2BASE-TL do with respect to the standards they reference. Generally speaking, everything that is not Ethernet-like is removed, or at least placed outside the scope of the specification (e.g., dual latency, ATM/STM support, NTR).

Frequency Plan

To meet the objective of offering 2 Mbps full-duplex at 2700 m, an ADSL system would have to be configured in the "Annex J" frequency plan.

---

[12] 2PASS-TL references ADSL2 because the original ADSL Recommendation did not contain a sufficiently symmetric frequency plan (Annex J).

This plan has the most symmetric upstream-downstream ratio (64 tones upstream vs. 192 tones downstream) of the different ADSL plans. The plan is still inherently asymmetric, but since the high-frequency tones are the first ones to go as the loop length increases, due to the increasing attenuation and crosstalk, there is a certain range of reaches within which the system's actual capacity is nearly symmetric.

The reliance on the "Annex J" frequency plan is what killed the 2PASS-TL proposal. It is well known that combining ADSL systems with different upstream-downstream split frequencies in the same cable bundle introduces a high level of NEXT in the overlapping portions of the spectrum. For this reason, "Annex B" and "Annex J" should not be deployed in bundles that already serve "Annex A" and/or "Annex I" customers, at the risk of a serious reduction in data capacity. In North America, where ISDN deployments are marginal, "Annex A" ADSL is the only type deployed (besides the pre-ITU T1.413 ADSL, which uses the same frequency plan). As a result, North American operators were very reluctant to allow an "Annex J" ADSL into their networks, and this reluctance caused the 2PASS-TL effort to fail.

It should be noted, however, is that SHDSL and Enhanced SHDSL pose the same spectral compatibility problems as "Annex J" ADSL. As SHDSL and Enhanced SHDSL are perceived as nonresidential DSL-types, they are somehow viewed as separate from the residential "Annex A" ADSL binders. In reality, this argument should work just as well for "Annex J" ADSL; it is not primarily a residential DSL-type, and may therefore be expected not to exist in binders that contain "Annex A" ADSL. From a technical point of view, all of the above can be shown to be spectrally compatible with "Annex A" ADSL up to a certain reach (deployment guideline) in accordance with spectrum management standard T1.417. From a practical point of view, the long-reach EFM Copper PHY was positioned as a business solution, and—like it or not—2BASE-TL was considered the best fit.

6

Point-to-Point Fiber

*"Point-to-point is the topology Ethernet has
used successfully for over a decade."*
—Pat Kelly[1]

Introduction to Point-to-Point Fiber

Optical Communication Basics

The physics of fibers. Optical communication is based on transmission of lightwaves through thin strands of glass or plastic fiber (there are also various wireless forms of optical communication, but these are not addressed by the Ethernet in the First Mile standard). As is the case with copper pairs, the fiber acts as a guiding medium for the electromagnetic wave onto which information is modulated. The main difference with the twisted copper pairs is the frequency of the electromagnetic waves: Whereas copper supports frequencies up to the multi-megahertz ranges, fiber can only be used for the terahertz waves that make up light (including nonvisible infrared light). The extremely high frequency of these waves makes them very useful for use as carrier waves, supporting very high modulation rates and hence very fast data transmission.

The principle of the transmission of light through multi-mode optical fiber (MMF) is best understood by taking the high-frequency approximation, which describes light as rays. The fiber consists of a core with a high refractive index n_1 (typical diameter values are 50 μm

---

[1] EFM Tutorial Session, July 2001.

and 62.5 µm), surrounded by a cladding with a lower refractive index n_2 (125 µm in diameter), most often surrounded in turn by a jacket or coating to protect the fiber from mechanical damage. The physics at the core/cladding interface cause a phenomenon known as *total internal reflection* for all incident waves that are at a launch angle (the angle between the direction of incidence and an axis orthogonal to the interface plane) larger than

$$\theta = \sin^{-1}\frac{n_2}{n_1} \qquad (6.1)$$

This means that light travelling along the length axis of the fiber strand, or at an angle smaller than

$$\frac{\pi}{2} - \theta \qquad (6.2)$$

with respect to that axis, remains confined inside the core.[2]

A simple glass/air interface would exhibit the same physical phenomenon of total internal reflection. However, the large difference in refractive index between glass and air leads to a very small angle θ, which means that light travelling at a great variety of angles with respect to the long axis will undergo total internal reflection and remain confined inside the fiber. Light bouncing between the sidewalls of the fiber at different angles will travel different distances (and hence, different times) between the two ends of the fiber, leading to the phenomenon of *multipath dispersion*. This phenomenon smears out short lightpulses in time, causing inter-symbol interference (ISI) and making the medium less useful for high-bitrate communications. For this reason, fibers that are used have a core and a cladding with slightly different refractive indices (Δ_n = 1% is a typical value). The angle at which total internal reflection occurs is thus kept small, such that only small variations in propagation angle are sustained within the fiber.

The variations in refractive index are obtained by "doping" the material with the appropriate impurities during the production process. Instead of having a radial variation of the refractive index, which looks like a step function (step index fiber), the fiber may also exhibit a gradual radial change in the refractive index. It can be shown that all rays

---

[2] Incidentally, it also means that light from outside cannot break into the fiber and propagate inside the core.

travelling on a confined path through a fiber with a parabolic index profile undergo the same propagation delay between any two fixed points. Hence, fibers with a parabolic index profile do not suffer from multipath time dispersion.

A more detailed analysis, which uses Maxwell's equations instead of the ray model, reveals that the different independent solutions of these equations (called *modes*) can have different propagation speeds. The dispersion associated with this variation in speed is called *intermode dispersion* and is roughly equivalent with the multipath dispersion in the ray model. Reducing the diameter of the fiber core limits the number of electromagnetic modes in which the light can propagate. Fiber designed to support only a single mode of propagation is called single-mode fiber (SMF). These fibers are produced with a much thinner core (e.g., 9 μm). Although slightly more expensive to deploy than multi-mode fiber, SMF has the advantage of being able to operate over longer distances, because it does not suffer from multi-mode interference.[3]

A different kind of dispersion is *material dispersion*. This is an effect of the wavelength-dependence of the refractive index of the fiber material. A pulse of light that is not perfectly monochromatic will be broadened by this effect.

Attenuation in fiber optical systems is caused by interaction of the photons with the fiber material itself, or with impurities in the fiber. Typical values of attenuation are in the order of 0.3 dB/km (1550 nm window) to 0.5 dB/km (1310 nm window).

Lasers and photodiodes.   As all the data and signal processing inside the point-to-point fiber transceiver happens in *electronic* circuits, there has to be a point where electrical signals are converted into optical signals and vice versa. The former conversion is performed by a laser (alternatively, light-emitting diodes (LEDs) can be used in short-reach applications); the latter by a photodiode. Apart from these conversions, optical transceivers are very similar to other kinds of transceivers.

Unlike the EFM Copper Physical Layer entity sublayers (PHYs), most optical PHYs (including EFM over point-to-point fiber and EPON) represent data on the medium in a purely binary way, known as on/off transmission. In *non-return to zero* (NRZ) coding, a "1" is represented by a pulse of light, a "0" is represented by a dark period; *non-return to zero inverted* (NRZI) inverts the signal on a "1" and leaves the

---

[3] Strictly speaking, single-mode fiber supports two degenerate modes corresponding to different polarizations. These modes suffer a small *polarization mode dispersion*, which is practically insignificant at rates up to 10 Gbps.

signal unchanged for a "0." Even though sufficient margin may be available to support more complex multi-level linecodes, this is uncommon in present-day optical systems. The extra margin is used to increase the reach or to improve the bit error ratio (BER) of the system. When higher bitrates are required, it is easier to crank up the symbol rate than to use multi-level transmitters and detectors. Nonbinary modulation is used in certain specialized domains, such as video distribution in hybrid fiber-coax networks (using AM and FM for analog transmission, and QAM for digital transmission).

For the reader unfamiliar with the principles of opto-electronics, a very brief overview is provided here. As this subject is very broad and complex, and as it is based on the even broader subject of semiconductor physics, this account is simplified and by no means complete. The main goal is to expose the issues that impact the cost and performance of an optical EFM system.

All light-emitting devices are based on the same quantum mechanical principle according to which an atom with an electron in an excited state emits a photon (quantum of light) when it falls back to a lower-energy state, the frequency ν of the light being proportional to the difference E between the energy levels ($E = h\nu$, where $h = 6.626068 \times 10^{-34}$ m^2 kg s^{-1} is Planck's constant). Conversely, an atom can absorb a photon by placing an electron into a state that is at an energy level that is E above its original state.

Within the atomic lattice of a semiconductor, the energy levels of electrons are no longer governed by the properties of the atoms alone, but also by the characteristics of the lattice; specifically, the discrete energy levels of electrons will become energy bands. The interesting phenomena take place in the "band gap" between the highest energy band that is normally populated with electrons (the valence band) and the lowest energy band that is normally devoid of electrons (the conduction band). Emission occurs when an electron "drops" from the conduction band into an empty spot (a hole) in the valence band, getting rid of a quantum of energy (a photon) in the process. Absorption consists of an incoming photon causing an electron to "jump up" from the valence band into the conduction band, leaving a hole in the valence band. Both the electron in the conduction band and the hole in the valence band are inclined to move under the influence of an electric field and are thus the individual carriers of *electrical current* in the semiconductor.

The width of the band gap E_g now determines the frequency ν and the wavelength λ of emitted or absorbed photons ($\nu \geq E_g/h$ or $\lambda \leq hc/E_g$). The net result is that semiconductor devices can be built that convert electric current into photons of a particular frequency and vice versa. The different band gaps of different semiconducting materials explain the preference of specific materials for specific optical applications.

Whereas the emission of photons in a LED depends on individual uncorrelated electron-hole recombinations, lasers depend on the phenomenon of *stimulated emission.* For stimulated emission to happen, three factors are required. First, amplification is achieved by electrically "pumping" electrons to the excited state. Second, spatial and temporal coherence is obtained by terminating the laser cavity with "mirrors"—zones with high reflectivity within the semiconductor—which give a standing wave nature to the photon field inside the cavity, causing a resonance effect. Finally, a "population inversion" must occur: An excess of electrons in excited states is waiting to emit photons upon interaction with the photons already present in the laser cavity—this requires a sufficiently high electrical current. The resulting high-power coherent light source is particularly suited for efficient optical fiber communication, thanks to the high level of optical power and its coupling into the fiber, the multi GHz-speed of power level modulation, and the narrow spectral width.

In both LEDs and lasers, the optical output power can be modulated by varying the input current. Unfortunately, changes in temperature strongly affect the properties of all semiconductor devices, including opto-electronic devices. Hence, some form of temperature control or temperature compensation is required to get a predictable optical output power.

The emission wavelength of a Gallium Arsenide (GaAs) device is 870 nm; for longer wavelengths (e.g., 1300 nm or 1550 nm), InGaAsP devices can be used. Fabry-Perot lasers are multi-longitudinal mode lasers, causing a relatively high mode partition noise (MPN) and dispersion, which limit their capacity. To decrease the MPN, distributed feedback (DFB) lasers can be used. The signal from a single-frequency laser will suffer a certain amount of interference with its reflection at the receiver, which should be taken into account as an additional noise term.

Unlike traditional lasers, Vertical Cavity Surface Emitting Lasers (VCSELs) are vertical structures (where "horizontal" is to be understood as the plane of the semiconductor wafer, and "vertical" is the perpendicular direction). The light-emitting surface is at the top of the structure, which is accessible immediately after wafer production (even without cutting the wafer first), which simplifies yield testing. An additional advantage is the possibility to create dense two-dimensional

VCSEL arrays. Although VCSELs are a relatively recent technology, their current consumption properties and packaging advantages give them a wide range of potential applications.

Photodiodes are diodes that are designed to efficiently absorb photons and generate electrical current. They do not provide a perfect conversion of optical power into electrical current; there are limitations to the bandwidth and dynamic range of the photodiode, and a certain amount of noise will be added to the signal. Furthermore, a very small "dark current" will be flowing even when no light is being detected, which introduces additional noise.

As photonic absorption primarily takes place in the depletion zone,[4] the efficiency of photodiodes can be influenced by controlling the width of the depletion zone in the manufacturing process. This can be done by adding an "intrinsic" zone (semiconductor material that is not doped either way) between the p-type zone and the n-type zone; the depletion zone will extend roughly over the width of the intrinsic zone. These diodes are usually referred to as "PIN-diodes."

In an optical receiver, the photodiode that acts as a detector is followed by an electronic amplifier and signal processing logic that converts the electronic signal to the required format. The digital signal processing stage includes clock detection and decision of the individual bits as "0" or "1." Certain photodiodes provide internal amplification by applying a strong reverse bias field, which causes the carriers to acquire enough energy to generate new electron-hole pairs (avalanche ionization). These photodiodes are known as "avalanche photo diodes" (APDs).

The EFM standard does not explicitly mandate the use of any particular kind of lasers or photodiodes, but the specified optical parameters will be more easily met with some types of components than with others. As component manufacturing technology evolves, and as the adoption of different standards changes the market dynamics of components meeting certain requirements, solutions that are not economically feasible today may become cost-effective tomorrow. For this reason, I will avoid making any direct recommendations as to the use of a particular optical technology in EFM implementations.

---

[4] This is the zone around the junction between the p-type semiconductor (doped to have an excess of holes) and the n-type semiconductor (doped to have an excess of electrons). The depletion zone has a reduced number of free carriers, due to diffusion of free carriers across the junction and recombination with oppositely charged carriers.

Design Choices for Optical Communications Systems

Attenuation and dispersion are the main factors that limit the data capacity of a fiber channel. The design of a fiber-based transmission system is therefore an exercise in budgeting the attenuation and the dispersion over a given length of fiber between the transmitter and the receiver. The receiver needs to be sufficiently sensitive to be able to decode a signal whose average power is equal to the transmitter's launch power minus the loss introduced by the connectors, the transmission through the fiber, splitters (if present), and any other predictable impairments.

For fused quartz, a common material for optical fibers, the material dispersion reaches a minimum at a light wavelength of 1300 nm (around 6.4 ps/km-nm.)[5] Fiber attenuation is at its lowest (down to 0.15 dB/km) around 1550 nm, with another interesting local minimum (down to 0.4 dB/km) around 1300 nm. In practice, 1550 nm and 1310 nm are the frequencies most commonly used for long-reach applications.

The actual transmit frequencies will not exactly coincide with these numbers, but they will be within a frequency "window," the width of which is limited by the occurrence of OH^+ ion absorption peaks in the spectrum, notably the one at 1400 nm. The common windows and their band designations are listed in Table 6.1.

It is no coincidence that EFM PHYs over bidirectional point-to-point fiber operate at or around these frequencies in downstream and upstream, respectively. The dual-fiber LX10 solutions operate at 1310 nm in both directions. In both cases, the use of SMF significantly reduces the total dispersion, thus further increasing the reach of the system.

TABLE 6.1   Optical Transmission Bands

Band	Wavelength (nm)
	820–900
O	1260–1360
E	1360–1460
S	1460–1530
L	1530–1565
C	1565–1625
U	1625–1675

---

[5] The dispersion coefficient should be read as a maximum amount of temporal spreading (in ps) per km of fiber length and per nm of spectral width.

For short-reach applications, attenuation and dispersion are less stringent problems. One can build fiber-optical systems using any wavelength in the visible and infrared spectrum up to approximately 1800 nm. GaAs-active lasers operating at frequencies in the window around 850 nm are popular for short-reach applications, because they are relatively easy and cheap to produce.

For a given bitrate, transmit power, and BER, the reach of an optical transmission system can be increased by specifying the optical components more tightly: The impact of different component properties is discussed in the sections below. These reach improvements obviously come at a certain cost. Ethernet has traditionally chosen to remain on the cheaper side of the cost/reach trade-off; as a result, more expensive commercial components are available that go further than the reach specified by 802.3, while remaining interoperable with standard components. The nominal BER supported by all optical EFM PHYs is 10^{-12}. This is a very stringent requirement, compared to some other commercially available optical transmission systems (e.g., GPON has a nominal BER of 10^{-10}).

For half-duplex operation, mentioned here for historical reasons only, the maximum reach is limited by a completely different consideration: In order for the CSMA/CD protocol to work, a locally detected collision must be signaled back to the remote end sufficiently and quickly. The round-trip time of the channel (in bit times at the given bitrate) therefore limits the maximum usable network reach to values that are often far below the physical reach of the PHY.

The EFM Task Force used the 10Gigabit Ethernet Link Model which was maintained and extended by David Cunningham and Piers Dawe [16] to select the parameters for its optics specifications. Each of these parameters has its impact on the reach of the optical PHYs and on the available choice of optics technology (and hence, on component cost). The model supports a generalized power budget calculation, which includes the classical power budget assessment (transmit power, insertion loss, receiver sensitivity) as well as time-domain impairments (jitter, dispersion) that are converted to an equivalent statistical power penalty. The power and time dimensions of the modeled optical system can also be visualized as a two-dimensional "eye diagram" (e.g., Figure 6.2).

The eye diagram is a well-known and important tool in transmission engineering: It is an oscilloscope measurement (or a computer simulation) of a large number of transmitted symbols, visually superimposed, the symbol transitions forming the characteristic eye shape that gives this diagram its name. The diagram reveals information about the signal,s tolerance for noise (vertical eye opening) and jitter (horizontal eye opening). The practical use of eye diagrams in testing Gigabit Ethernet is concisely described in [17].

Optical communication standards, such as EFM, specify masks to which the measured transmit eye diagram of a system must comply in order to meet the requirements of the standard. Compliance requires that the opening of the measured eye diagram remains greater than the zone defined by the mask during normal system operation, allowing for a small number of statistical mask violations (5×10^{-5} hits per sample in EFM optics).

The 10Gigabit Ethernet Link Model is implemented as an open-source Excel-based tool that is available on the EFM website. It uses generally accepted values for the physical constants in the equations and the following parameters to model the behavior of the system.

Choice of fiber.   Unlike the other optical parameters, the choice of fiber is a selection from a discrete set of options. Different fiber standards exist, specifying maximum attenuation rates, leading to a given channel insertion loss *ChIL* at a given reach. SMF, with its lower dispersion penalty, is an obvious choice in a context where reach matters. The attenuation of a fiber can be influenced by the manufacturing process (e.g., dehydrogenated fiber). The choice of fiber is taken into account by the Link Model in the formulas for dispersion and attenuation.

Transmit power.   The transmit power is specified as the optical modulation amplitude P_{TXOMA}, i.e., the difference between the power level of a "1" and the power level of a "0." The "0" power level is P_{er} [dB] lower than lower than the "1" power level, a parameter known as the extinction ratio. From these two parameters, the *average launch power* can be calculated, which is also sometimes used for modeling purposes. As the transmit power is the starting point for the calculation of the link budget, increasing it has a direct impact on the reach of the optical system. The practical maximum transmit power for semiconductor lasers is in the order of 10 dBm; other, more powerful types of lasers do not have the efficiency, size, and integration benefits that semiconductors offer.

Relative intensity noise.   Fluctuations in the output of the transmitter laser are a noise factor known as relative intensity noise (RIN).

Bandwidth.   The bandwidth of the modulated signal is a function of the modulation rate B—fixed for a given bitrate and linecode—and more specifically of the rise time t_s of the power level transitions (the time that elapses between reaching 20 percent of the power level and reaching 80 percent of the power level). The bandwidth contributes to the dispersion-related penalties.

Center wavelength. As explained above, center wavelengths λ_C of 1310 nm and 1550 nm provide the best dispersion and attenuation properties.

Spectral width. The spectral width σ_λ of the optical source expresses the relative amount of power at wavelengths around the center wavelength; this parameter also contributes to the dispersion related penalties. A smaller spectral width can only be obtained by using a more precise, more expensive laser (e.g., DFB laser).

Deterministic jitter and duty cycle cistortion. The transmitter-induced variation of the timing of individual bits relative to a fixed clock is known as the deterministic jitter (DJ). Duty Cycle Distortion (DCD) is a related effect measured at the end of the bit symbol. These are a matter of how swiftly the laser reacts to changes.

Receiver sensitivity. The parameter S_{OMA}, also specified as an optical modulation amplitude, determines the lowest signal power that can reliably be detected at the receiver. It depends on the absorption rate in the photodiode; detectors with internal gain (avalanche diodes) provide the best sensitivity. The reach of a fiber system cannot be extended indefinitely by increasing the receiver sensitivity, because the thermal noise and bandwidth-dependent impairments will take over at some point.

The link budget method is illustrated in Figure 6.1. By feeding the EFM media requirements (single or dual SMF) and reach requirements (10 km for point-to-point fiber; 10 or 20 km for EPON) into the model, and taking into account the cost impact of the use of different kinds of components, the parameters listed in the remainder of this chapter and the next were selected for the optical PHYs in the EFM standard.

In real-world deployments, the link budget is degraded by factors such as splices, connectors, bends in the fiber, and contamination of the optical surfaces, which are accounted for in the link model. Additionally, a certain operational margin M must be observed in the link budget method for unexpected losses.

P_{TXOMA}

$ChIL + P_{mpn} + P_r + P_{rin} + \dfrac{P_c}{2} + P_{mn} + M$

$P_{er} - 3dB$

$<P_{TX}>$

Power
Budget (dB)

Power
Budget (dB)

SRS_{OMA}

$\dfrac{P_c}{2} + P_{isi}$

S_{OMA}

$P_{er} - 3dB$

$<P_S>$

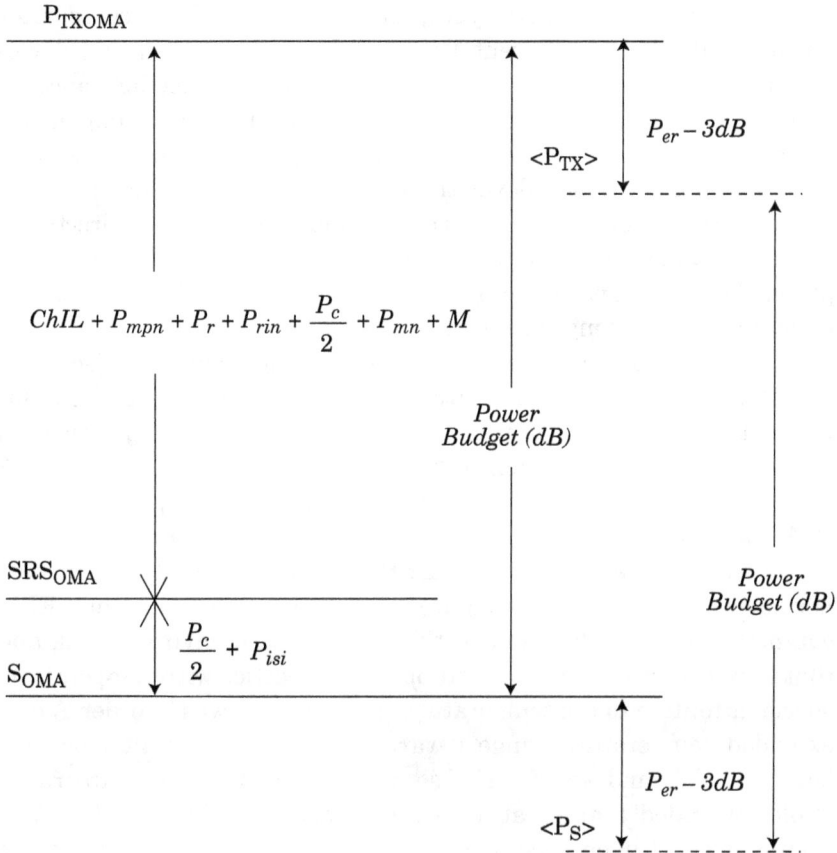

Figure 6.1   Relation of the variables in the link budget calculation (source: 10Gigabit Ethernet Link Model [16]. Courtesy of David Cunningham and Piers Dawe.)

EFM Innovations

In addition to extending the reach of Fast Ethernet and Gigabit Ethernet to 10 km, the EFM Task Force introduced three major innovations with respect to the existing fiber-optic Ethernet PHYs: support for bidirectional operation over a single fiber[6] (the BX10 PHY family), 100 Mbps PHYs operating SMF (100BASE-LX10/BX10), and specifications for components in extended temperature ranges.

---

[6] Single fiber is sometimes referred to as "simplex" fiber, as opposed to "duplex" fiber for two strands. These terms are not to be confused with the terms "simplex," "half duplex," and "full duplex" operation, which indicate, respectively, whether a given PHY/medium combination supports transmission in one direction, both directions alternatingly, or both directions simultaneously.

For the single-fiber port types, the EFM Task Force chose after much debate to use different wavelengths for upstream and downstream transmission. This eliminates the need for reflection cancellation and makes the detection of fiber disconnect slightly simpler. On the other hand, it introduces the need for two different transceiver subtypes (upstream and downstream) for each speed, which can only communicate when combined correctly. Dual-wavelength bidirectional optical modules must contain a special wavelength division multiplexer (WDM) to divert the incoming wavelength to the receiver and the outgoing wavelength to the fiber.

Specifying 100BASE-X over SMF was not an "original" objective of the EFM Task Force. At the March 2002 meeting, a successful Call for Interest was held for a project to specify a 100 Mbps PHY over dual SMF. At the closing session of that same meeting, the IEEE 802.3 Working Group adopted this topic as an additional objective for the EFM Task Force.

The current specifications in IEEE Std. 802.3-2002 do not include normative environmental requirements. Given the fact that EFM equipment may be deployed in thermally hostile environments, and that uncooled lasers change their optical properties with temperature, special attention to the temperature specifications was in order. A first extended temperature range ("warm extended") aims at operation between –5°C and +85°C. The second extended temperature range ("cold extended") aims at operation between –40°C and +60°C. Compliance with one, both, or none of the extended temperature ranges may optionally be declared by all optical EFM PHYs.

100 Mb/s Family

Introduction to 100BASE-X PCS and PMA

The 100BASE-X PHY family, which was added to IEEE Std. 802.3 in 1995 by the 802.3u "Fast Ethernet" project (the second phase of which was also chaired by Howard Frazier), is largely based on earlier standards for Fibre Distributed Data Interface (FDDI, ANSI X3.263-1995). The name 100BASE-X refers to the Physical Coding Sublayer (PCS), which is common between several 100 Mbps PHYs (originally 100BASE-TX and 100BASE-FX; since the advent of EFM, this also includes 100BASE-LX10 and 100BASE-BX10).

The new 100 Mbps PHYs added by EFM use the existing 100BASE-X PCS, as specified in Clause 24 of IEEE Standard 802.3, with one enhancement for subscriber access networks: The PCS is now capable of transmitting data regardless of the value of the *link_status* variable. This means that a unidirectional link can exist, which may be useful for troubleshooting remote faults (see Chapter 8).

The 100BASE-X PCS encodes Media Independent Interface (MII) data nibbles into five-bit code-groups (4B/5B), which are transmitted serially over the medium. The PCS takes care of serialization/deserialization of these codegroups at the interface with the underlying Physical Medium Attachment Sublayer (PMA) and the appropriate translation of the MII signals for transmit, receive, carrier sense, and collision detect to and from PMA signals.

The 5-bit code-group space provides sufficient possible codewords (32) to accommodate the 16 possible data nibbles (see Table 6.2) and necessary control code-groups. The control code-groups are known as /I/, /J/, /K/, /T/, /R/, and /H/ (symbol /V/ is used for undefined code-groups). The /I/ code-group (IDLE, equal to all ones) is used as an inter-stream fill code. A sequence /J/K/ (binary 11000 10001) is used to signal the start of a data stream, and a sequence /T/R/ (binary 01101 00111) is used to signal the end of a data stream. Code-group /H/ (binary 00100) is used as a deliberate transmit error.

As a result, a MAC frame will be converted to a stream consisting of a Start-of-Stream Delimiter (SSD) /J/K/, the remaining 6 octets of the preamble, the SFD, the data frame, an End-of-Stream Delimiter (ESD), and a number of idle codegroups /I/ filling the remainder of the inter-packet gap.

TABLE 6.2   Data Encoding in 4B/5B

Data Nibble	Binary Encoding [4:0]
0	11110
1	01001
2	10100
3	10101
4	01010
5	01011
6	01110
7	01111
8	10010
9	10011

(continued)

TABLE 6.2   Data Encoding in 4B/5B *(Continued)*

Data Nibble	Binary Encoding [4:0]
A	10110
B	10111
C	11010
D	11011
E	11100
F	11101

100 Mbps Optics

As with all 100BASE-X signalling systems, 100BASE-LX10 and 100BASE-BX10 operate at a signaling speed of 125 Mbaud (±50 ppm). The "L" indicates that long wavelengths are used, that is, 1260–1360 nm (100BASE-LX10), 1260 nm–1360 nm (100BASE-BX10-U), and 1480 nm–1580 nm (100BASE-BX10-D).

As 100BASE-LX10 and 1000BASE-BX10 are essentially intended for fiber-to-the-home deployments, low cost was deemed very important; after all, this technology would have to compete with 10PASS-TS and 100BASE-TX from the cabinet or basement, both capable of offering a similar bitrate over a cheaper medium. The optical parameters of 100BASE-LX10 and 100BASE-BX10 were therefore chosen to be within the operational range of Fabry-Perot lasers.

TABLE 6.3   EFM vs. the Existing 100 Mbps Optical Ethernet PHY

Port Type	Medium	Wavelength	FD Reach	Avg. Launch Pwr.	RX Sensitivity
100BASE-FX	2 MMF	1300 nm	2 km	−20 dBm / −14 dBm	−31 dBm
100BASE-LX10	2 SMF	1310 nm	10 km	−15 dBm / −8 dBm	−25 dBm
100BASE-BX10-U	1 SMF	1310 nm	10 km	−14 dBm / −8 dBm	−28.2 dBm
100BASE-BX10-D	1 SMF	1550 nm	10 km	−14 dBm / −8 dBm	−28.2 dBm

100BASE-LX10.   100BASE-LX10 is based on the SONET OC-3 IR-1/SDH STM-1 S-1.1 standard with appropriate modifications and offers a service interface that is identical to the existing multi-mode

Tx Eye Diagram (No Noise)
Solid: Test Rx   Dashed: Target Link and Rx

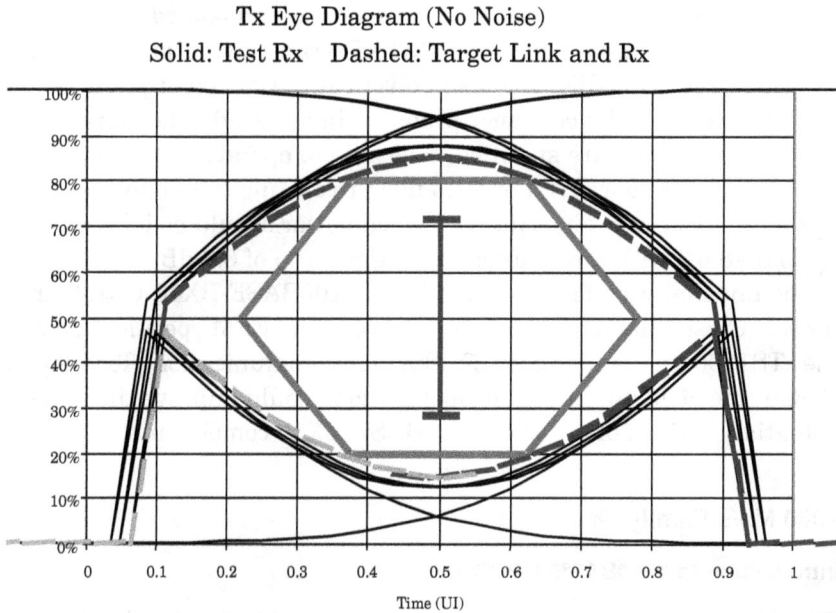

Figure 6.2   Eye diagram for 100BASE-LX10. Source: 10Gigabit Ethernet Link Model [15]. Courtesy of David Cunningham and Piers Dawe.

standard 100BASE-FX. This port type uses two strands of SMF to achieve full-duplex operation. The transceiver is specified to be able to operate at reaches of up to 10 km, on specific types of SMF (IEC60793-2 Categories B1.1 and B1.3), tolerating a channel insertion loss of up to 6 dB. The average launch power should be between –15 dBm and –8 dBm, with a minimum extinction ratio of 5 dB.

Figure 6.2 shows a simulated eye diagram for the optical parameters associated with 100BASE-LX10, simulating 10 km of SMF. The diagram was generated with the 10Gigabit Ethernet Link Model, which was also used by the EFM Optics Sub Task Force to evaluate different proposals for optical parameters.

100BASE-BX10-D/U.   Contrary to 100BASE-LX10, 100BASE-BX10 operates over a single strand of fiber. Two different transceiver types are specified, one for each side of the link: 100BASE-BX10-D and 100BASE-BX10-U. The transmission direction from the operator network towards the subscriber is typically denoted as "downstream" (D), and the opposite direction is typically denoted as "upstream" (U). The optical parameters have been selected such that the requirements for the 100BASE-BX10-U are less stringent, such that they can be met with slightly cheaper optics. It is therefore recommended, though not mandatory, to employ 100BASE-BX10-D transceivers at the central office and 100BASE-BX10-U transceivers at the customer premises (or the site closest to the customer premises in a cascaded network).

The transceivers are specified to be able to operate at reaches of up to 10 km on Category B1.1 or B1.3 fiber, tolerating a channel insertion loss of up to 5.5 dB. The average launch power should be between −14 dB and −8 dB with a minimum extinction ratio of 6.6 dB.

The unusual wavelength range for the 100BASE-BX10-U transmitter was chosen to increase overlap between the EFM specification and the TS1000 specs defined by the Telecommunication Technology Committee of Japan.[7] Such harmonization should improve the market potential and economics of the 100BASE-BX10 components.

1000 Mb/s Family

Introduction to 1000BASE-X PCS and PMA

The Gigabit Ethernet project (IEEE 802.3z) adds another PCS to the list of those designated by an "X" in the port type name, coincidentally all based on existing ANSI X3 standards (1000BASE-X is based on X3.230 Fibre Channel standard). The new 1000 Mbps PHYs added by EFM use the existing 100BASE-X PCS, as specified in Clause 36 of IEEE Standard 802.3, with one enhancement for subscriber access networks: The PCS is now capable of transmitting data regardless of the value of the *link_status* variable. This means that a unidirectional link can exist, which may be useful for troubleshooting remote faults (see Chapter 8).

The 1000BASE-X PCS accepts data from the Gigabit Media Independent Interface (GMII). This sublayer is responsible for the encoding of data into 10-bit code-groups (8B/10B). It also converts information from the PMA to generate the CRS and COL signals on the GMII, and manages the auto-negotiation process.

The 10-bit code-group space provides sufficient possible codewords (1024) to accommodate two representations for each of the 256 possible data octets and twelve special code-groups used for control purposes (/K28.0/ through /K28.7/, /K23.7/, /K27.7/, /K29.7/, and /K30.7/). The control messages are sent as eight "ordered_sets" consisting of a

---

[7] The Telecommunication Technology Committee produced an English translation of its draft specification for TS-1000 "Optical Subscriber Line Interface—100 Mbit/s Single-fiber Bi-directional Interface by WDM" and made it available to the EFM Task Force.

special code-group, followed by up to three data code-groups known as Configuration (/C1/ and /C2/), Idle (/I1/ and /I2/), Carrier_Extend (/R/), Start_of_Packet (/S/), End_of_Packet (/T/), and Error_Propagation (/V/).

For each of the symbols to be transmitted, two encodings are provided. Depending on the running disparity, which is calculated as a function of the number of ones and zeros in the previously transmitted symbols, the RD- or RD+ version of the symbol is selected for transmission, in order to keep the average disparity neutral. This improves the timing recovery at the receiver side. Additionally, there is the notion of "comma": This is a sequence of binary symbols (0011111 or 1100000), which can only occur as part of /K28.1/, /K28.5/, and /K28.7/. As a result, commas occur if and only if a /C/ or /I/ ordered_set is being transmitted. Since these are sent continuously between packets, commas can be used to obtain symbol delineation in the lower layers of the receiver.

Transceiver Modules

A number of industry-initiated specifications or "multi-source agreements" (MSAs) exist for pluggable transceiver modules. The interfaces specified in these agreements are not defined in IEEE Std. 802.3, but they are used as *de facto* standards, because they are supported (by design) by a large number of equipment producers.

In principle, the physical interface between the component responsible for data-related operations and the component responsible for medium-related operations can coincide with any functional interface provided explicitly or implicitly by the standard; this is precisely the great advantage of defining sublayers. IEEE Std. 802.3 has a tradition of explicitly specifying compliance points for at least the interface between the MAC and the PHY (MII, GMII, and XGMII; all optional) and the interface between the PMD and the medium (MDI). Other physical interfaces specified by the standard include the 1 Mbps and 10 Mbps AUI, and the 10 Gbps XAUI (extension of the XGMII).

Nothing prevents individual vendors or groups of vendors from implementing a PHY as a group of components or modules that connect to one another at other functional interfaces than the ones explicitly specified in the standard. The following sections provide a few examples of such an approach.

Gigabit interface converter.   The gigabit interface converter (GBIC) is a "removable serial transceiver module designed to provide gigabaud capability for Fibre Channel (FC) and other protocols that use the same physical layer." The specification, originally published in 1995 (version 1.0) and finalized in 1998 (version 4.5), consists of electronic, electrical, and physical interfaces. Later versions contain clarifications of the base document and technology-specific additions. The term transceiver—again, not official 802.3 terminology—is to be understood as "an electro-optical subsystem whose function is to convert optical signals to electrical signals and vice versa" [44]. Functionally, the GBIC contains the PMD portion of a 1000BASE-X Gigabit Ethernet PHY (in the case of 1000BASE-T; the GBIC module contains some additional conversion functions).

GBIC-compatible interfaces are common in Gigabit Ethernet systems. While the manufacturer of the system focuses on the data parsing and switching functions, the user of the equipment can extend the system with GBIC transceivers meeting specific media (both copper and fiber) and reach requirements. The MAC, 1000BASE-X PCS, and 1000BASE-X PMA would be part of the Gigabit Ethernet system (the "host"), while the PHY is implemented in the GBIC module. GBIC modules exist for the different 1000BASE-X PHYs as well as for 1000BASE-T.

The more recent Small Form-factor Pluggable (SFP) Transceiver is an alternative solution with similar benefits. The advantages of SFP over GBIC are its smaller form factor (hence the name; it is about 13.7 mm × 8.6 mm versus 30.48 mm × 12.01 mm for GBIC) and the lower operating voltage (3.3 V versus 5 V for GBIC, which reduces the power consumption of the host). The size of the module is of great importance to make cost-effective equipment with high port density.

Beyond 1 Gbps. For completeness, I will mention that similar industry standard modules exist for 10 Gbps applications.

Traditional *transponders*, such as those standardized by the XENPAK, XPAK, and X2 MSAs, interface directly with the Reconciliation Sublayer through a parallel interface (XGMII, or its reach-extended counterpart XAUI) and provide all PHY functions inside the module. To improve flexibility, the physical interface can also be placed a little further down the sublayer stack: A 10 Gigabit Ethernet PHY can, for example, be composed of a first component dealing with the PCS functions (64B/66B encapsulation) connected through an XSBI (16 × 644 Mbps) interface to a transponder containing a serializer/deserializer (serdes) and the necessary optics.

Finally, the physical interface can also correspond with the functional PMD/PMA interface. The serdes function is then integrated in the first component, which interfaces serially with an optical *transceiver* module such as the 10 Gigabit Small Form Factor Pluggable Module (XFP) created by the XFP MSA Group. XFP consists of "a specification for a module, cage hardware, and IC interfaces for a 10-Gbit hot pluggable module converting serial electrical signals to external serial optical or electrical signals." Hence, the XFP is the 10-Gbps equivalent of the GBIC and the SFP.

As the optical *transceiver* contains only minimal digital functionality, and as it connects to the rest of the PHY through a serial interface, the actual module can be made much smaller and more power-efficient than in the case of a *transponder*.

1000 Mbps Optics

As with all 1000BASE-X signalling systems, 1000BASE-LX10 and 1000BASE-BX10 operate at a signaling speed of 1.25 Gbaud (±100 ppm). The "L" indicates that long wavelengths are used, that is, 1260–1360 nm (1310 nm nominal for 1000BASE-LX10), 1260 nm–1360 nm (1000BASE-BX10-U), and 1480 nm–1600 nm (1000BASE-BX10-U).

TABLE 6.4   EFM versus Existing 1000 Mbps Optical Ethernet PHYs

Port Type	Medium	Wavelength	FD Reach	Avg. Launch Pwr	RX Sensitivity
1000BASE-SX	2 MMF	770 nm–860 nm	550 m	−9.5 dBm	−17 dBm
1000BASE-LX	2 MMF	1270 nm–1355 nm	5 km	−11.5 dBm/ −3 dBm	−19 dBm
1000BASE-LX10	2 SMF	1310 nm	10 km	−9 dBm / −3 dBm	−19.5 dBm
1000BASE-BX10-U	1 SMF	1310 nm	10 km	−9 dBm / −3 dBm	−19.5 dBm
1000BASE-BX10-D		1490 nm	10 km	−9 dBm / −3 dBm	−19.5 dBm

1000BASE-LX10. This port type also uses two strands of SMF to achieve full-duplex operation. The transceiver is specified to be able to operate at reaches of up to 10 km on Category B1.1 or B1.3 fiber, tolerating a maximum channel insertion loss of 6 dB, or up to 550 m on MMF (50 µm or 62 µm), tolerating a maximum channel insertion loss of 2.4 dB.

Although Table 6.4 indicates that the existing1000BASE-LX standard was limited to a reach of 5 km, there are implementations on the market (under a variety of names) that go a lot further. The parameters of 1000BASE-LX10 were chosen to be compatible with these existing pre-standard long-reach (>10 km) 1 Gb/s implementations.

1000BASE-BX10-D/U.   Like 100BASE-BX10, 1000BASE-BX10 operates over a single strand of fiber. Two different transceiver types are specified, one for each side of the link: 1000BASE-BX10-D and 1000BASE-BX10-U. The optical parameters have been selected such that the requirements for the 1000BASE-BX10-U are less stringent, and they can be met with slightly cheaper optics. It is recommended, though not mandatory, to employ 1000BASE-BX10-D transceivers at the central office and 1000BASE-BX10-U transceivers at the customer premises (or the site closest to the customer premises in a cascaded network). The transceiver is specified to be able to operate at reaches of up to 10 km on Category B1.1 or B1.3 fiber, tolerating a maximum channel insertion loss of 5.5 dB.

7

Point-to-Multipoint

"EPON systems are a highly attractive access solution because of cost and performance advantage, resulting from their nature as all-passive networks, point-to-multipoint architecture, and native Ethernet protocol."
—EFMA EPON Tutorial

Introduction to Passive Optical Networks

Passive Optical Networks (PONs) are based on optical links for which all of the general observations in the section "Optical Communications Basics" in Chapter 6 hold unchanged. A PON is a network in which a plurality of stations are connected to each other by means of a fiber-optical network that does not contain any active equipment; the different fiber segments to which the stations are attached are coupled together by means of a passive (not electrically powered) optical coupler or splitter; this is a device that splits optical power from one fiber into multiple fibers and vice versa.

In contrast, the term "active star" is used to designate a network consisting of a central piece of active equipment (e.g., an Ethernet switch) to which different stations are connected by means of a fiber link. In respect to the active star, a PON has the advantages of not requiring electrical power in the field (thus reducing maintenance cost) and of reducing the total distance of fiber that needs to be deployed for a given set of subscribers, assuming that the splitter can be placed closer to the subscribers than active equipment would be. The PON equipment at the Central Office (CO) terminates only a single fiber for a number of subscribers, using less space and requiring less fiber handling than the point-to-point equivalent.

Before the EFM project, the Ethernet standard already contained a specification for a 10 Mbps passive optical star (10BASE-FP), which supports up to 33 hosts sharing a (half-duplex) medium consisting of fibers connected to a passive central splitter unit, using the Carrier-Sense Multiple Access with Collision Detection (CSMA/CD) medium access method. The Ethernet PON (EPON) is fundamentally different from 10BASE-FP in that it assumes an architecture in which there is a single head-end (Optical Line Terminal or OLT) and up to 16 subscriber units[1] (Optical Network Units or ONUs), connected to each other through a passive splitter or a cascade of passive splitters (shown as "S" in Figure 7.1).

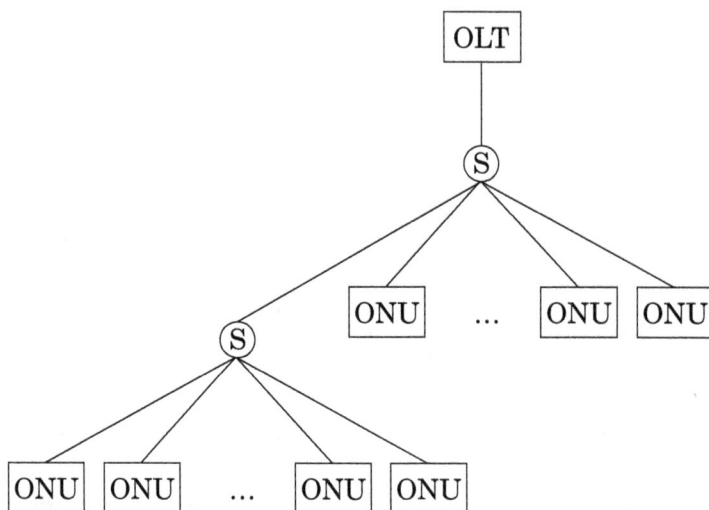

Figure 7.1   Example topology of an Ethernet passive optical network.

Downstream communication is broadcast in bursts at a rate of 1 Gbps from the OLT to the ONUs via the splitter, using Time Division Multiplexing (TDM). Upstream communication travels from each of the ONUs to the OLT in timeslots allocated by the OLT through the Multi-Point Control Protocol (MPCP), which is a form of Time Division Multiple Access (TDMA). Unlike downstream communication,

---

[1] The split ratio is not necessarily limited to 16; although this split ratio was the original objective, actual implementations may allow higher split ratios based on the optical link budget.

upstream communication in the EPON is not transmitted to the other ONUs. Upstream and downstream do not interfere with each other physically, as they use different optical wavelengths.[2]

EPON was introduced to the EFM project through an objective to *"support subscriber access network topologies: point to multipoint on optical fiber [...]."* This was later refined into two port type objectives:

- PHY for PON, $\geq$10 km, 1000 Mbps, single SM fiber, $\geq$1:16
- PHY for PON, $\geq$20 km, 1000 Mbps, single SM fiber, $\geq$1:16

The fact that there are "two" point-to-multipoint Physical Layer entity sublayer (PHY) specifications instead of one is another example of a Standards Development Organization (SDO) compromise required to get 75 percent approval for the proposal. The latter objective was added to have the best of two worlds: a cheaper shorter-reach solution (1000BASE-PX10) as well as a longer-reach solution, which is designed to serve a larger percentage of the market (1000BASE-PX20).

The downstream broadcast used in EPON makes it easy for subscribers to eavesdrop on transmissions addressed to other subscribers. Unlike their colleagues from Working Groups 802.11 (Wireless LAN) and 802.16 (Wireless MAN),[3] the 802.3 Working Group chose not to venture onto the treacherous grounds of privacy and security. Although the EFM Task Force at one point recommended adding "hooks for security and privacy" to the draft, the Working Group, remaining faithful to the strict layering dogma, which has proven so successful in the past, decided to let implementers of higher layers take care of higher-layer problems.

A task group on link security, spun off from the EFM EPON group, is currently active in IEEE Working Group 802.1. It will try to specify security mechanisms that can be used on any single IEEE 802 link, regardless of the MAC/PHY technology. This topic is further discussed in Chapter 9.

Just like the point-to-point Copper PHYs, EPON enters an arena previously occupied by Asynchronous Transfer Mode (ATM)-based technology. ITU-T has a complete ATM PON (APON) specification

---

[2] It would have been possible to give every ONU a different wavelength, creating a WDM PON. This approach was not selected for EPON, because of the large number of different transceiver types that would be needed in a WDM PON, and the resulting network management overhead.

[3] A good introduction to Wireless MAN and WiMAX can be found in [45].

(Recommendation G.983.1) for sub-Gigabit speeds (155/622 Mbps), extended to a complete Broadband PON system by a further set of Recommendations known as BPON. When the development of the EPON standard started, ITU-T had already commenced development of BPON's successor, the multi-protocol Gigabit PON (GPON) running at speeds up to 2488 Mbps. GPON supports both ATM and Ethernet transport.

Data Transfer

The EPON specifications consist of two port types (1000BASE-PX10 and 1000BASE-PX20), each with an OLT subtype and an ONU subtype.[4] Both port types contain the 1000BASE-X Physical Coding Sublayer (PCS), using 8B/10B coding (see the description of the 1000BASE-X PCS in the previous chapter).

The use of a symbol rate of 1.25 Gbaud combined with 8B/10B coding yields a net bit rate of exactly 1 Gbps, to be divided over the different EPON users. The use of this high-overhead PCS is sometimes mentioned as one of the major inefficiencies of the EPON standard with respect to the ITU-T's GPON Recommendation (together with the big frame size without support for fragmentation, the burst overhead, the transport of IPG, preamble, and SFD); on the other hand, the reuse of an existing coding scheme allows the reuse of existing IP blocks, thus improving the time-to-market and the total component cost.

As do all Ethernet port types, 1000BASE-PX10 and 1000BASE-PX20 connect to the existing CSMA/CD Media Access Control (MAC), but the actual media access method used by EPON bears very little resemblance to the original CSMA/CD philosophy. The main differentiation is introduced in the MAC Control sublayer (above the MAC) and the Reconciliation Sublayer (below the MAC), leaving the MAC almost entirely out of work.

The existing CSMA/CD protocol could have been used to resolve contention on EPON networks, as was done for 10BASE-FP, but this would have had serious drawbacks. Due to the point-to-point nature of the upstream links, ONUs would be unable to detect collisions of

---

4 The OLT and ONU have different protocol state machines, different optical requirements, and different economical constraints; it is therefore unlikely that a single box would offer both OLT and ONU capabilities. However, for the EPON SNMPv2 MIB, currently under definition in IETF, a writeable *dot3MpcpMode* object has been proposed, which would allow multi-mode EPON modems to be configured as an OLT or as an ONU.

frames originating from different ONUs. For the ONUs to be made aware of collisions, the OLT would have to transmit a jam signal downstream. With network diameters of up to 20 km, such a signal would require significant time to travel from the OLT to the ONUs, resulting in an unacceptable loss of efficiency [28].

Optics

Common Issues

Frequency plan. Both EPON types use the same frequency plan, with wavelengths in the 1480–1500 nm range (S-band) in downstream, and in the 1260–1360 nm range (O-band) in upstream, which is intentionally the same as the one in ITU-T G.984.2 GPON networks. This is important for the possibility to build combined GPON/EPON components (although a number of other differences between EPON and GPON complicate that matter), but also for the ease of deployment of (analog) overlay services such as broadcast TV, as is done today in BPON networks. The BPON standard specifies the use of the 1539–1565 nm band for additional digital services, or the 1550–1560 nm band for video distribution (both within the C-band).

Burst mode operation. The optics used in PONs need to be able to operate in "burst mode;" this means that transmit power in an ONU must be quickly turned on and off as a granted window opens and closes, and that OLT receivers must be capable of detecting and decoding signals that seem to hop from one power level to the next as different ONUs at different distances from the OLT take over from each other. At the ONU, the laser is turned on between bursts in response to the detection of data at the Gigabit Media Independent Interface (GMII). The PCS calls the PMD_SIGNAL.request(true) primitive, which is passed on from the PMA to the PMD. The PCS buffers the transmit data it receives from the GMII while the laser is turned on and allowed to stabilize. On completion of data transmission, when the PCS starts presenting idle codewords to the PMA service interface, PMD_SIGNAL.request(false) is called to turn the laser off. The data detector function and the resulting primitives are necessary to preserve layering in the EPON model; the functional effect that is achieved by this, is that the MAC gets to control the activation of the laser in the PMD, and that transmitter and receiver are stable by the time there is data to send.

Burst mode operation comes with the following basic dilemma: Relaxing the timing requirements on the receiver leads to simpler and cheaper receivers, but also to less efficiency and lower net bitrates. During the development of the EFM Draft, two forces further polarized the discussion on the timing issue in the Optics Sub Task Force. On the one hand, FSAN had already defined a set of strict timing requirements for its GPON specification (ITU-T Recommendation G.984.3). Lining up the EFM requirements with the FSAN specification could have stimulated low-cost mass-production of such components. On the other hand, existing optical Gigabit Ethernet components have less stringent requirements, and a market synergy with those components could be obtained by keeping the requirements loose.

This type of issue does not get resolved in a single meeting. The Task Force debated this issue at and between its November 2002 and January 2003 meetings. A gentle reminder by the optics editor that the value of making a choice would be greater than the cost we were trying to squeeze out of the silicon prepared the minds of the Task Force members for a choice at the Vancouver meeting. The looser specifications were selected (eventually specified as 512 ns laser on time; 512 ns laser off time).

1000BASE-PX10-D/U Specifics

The 1000BASE-PX10 PHY pair is specified to operate over distances of up to 10 km. The optical parameters of 1000BASE-PX10 were selected after modeling with the 10Gigabit Ethernet Link Model discussed in Chapter 6. In summary, the 1000BASE-PX10 PHY must tolerate a channel insertion loss of approximately 5 dB (taking a loss of 0.5 dB/km as a rule of thumb), in addition to the theoretical 12 dB associated with a 1:16 optical splitter and additional connector losses. Hence, the EFM standard prescribes maximum insertion losses of 20 dB (upstream) and 19.5 dB (downstream).

With these parameters, the 1000BASE-PX10 optics are very similar to the GPON Class A specification, which is designed for operation at reaches of up to 10 km with a 1:8 split factor. GPON Class A operates under a nominal optical loss between 5 dB and 20 dB.

1000BASE-PX20-D/U Specifics

The 1000BASE-PX20 PHY pair is specified to operate over distances of up to 20 km. The optical parameters of 1000BASE-PX20 were also selected after modeling with the 10Gigabit Ethernet Link Model. In summary, the 1000BASE-PX20 PHY must tolerate a channel insertion loss of approximately 10 dB, in addition to the theoretical 12 dB split factor and additional connector losses. Hence, the EFM standard pre-scribes maximum insertion losses of 24 dB (upstream) and 23.5 dB (downstream).

With these parameters, the 1000BASE-PX20 optics are very similar to the GPON Class B specification, which is designed for a maximum reach of 20 km or maximum 64 users (hence, a typical point of operation would be a PON reach of 10 km with a 1:32 split factor). GPON Class B operates under a nominal optical loss between 10 dB and 25 dB.

Note that there is no EPON equivalent to the GPON Class C optics. These are designed to operate at typical reaches of up to 20 km with a 1:64 split factor (15–30 dB loss). Given the particularly stringent requirements on the optics to achieve this performance, a 30 dB PHY would not be able to have the low cost typically associated with Ethernet PHYs. Whether the lack of a "Class C" PHY turns out to be a handicap in the competition with GPON remains to be seen. Nothing prevents individual vendors from manufacturing more performant equipment as the cost of opto-electronic devices evolves.

Forward Error Correction Sublayer

The optional Forward Error Correction sublayer (FEC) is a shim sub-layer between the PCS and the Physical Medium Attachment Sublayer (PMA). It receives packets from the PCS, performs FEC cod-ing on the data bytes in the packets, and transmits the packets with the /S_FEC/ delimiter and the additional parity information to the PMA. In the receive direction, the FEC performs octet alignment on the data coming in from the PMA, it detects and decodes the FEC par-ity code, and passes the corrected data up to the PCS, substituting idle codegroups /I/ for the parity bytes and error propagation codegroups /V/ for uncorrected blocks. The start and end symbols of FEC coded frames are constructed from existing 8B/10B codegroups, as shown in Table 7.1.

The specification of the FEC code can be found in ITU-T Recommendation G.975. It is a (255,239) Reed-Solomon code similar to the ones used in 10PASS-TS (see the section "Forward Error Correction" in Chapter 4). Prior to encoding, PCS data are divided in 239-symbol blocks, to which 16 parity symbols are added. If the final block in the packet is shorter than 239 symbols, it is padded with zeros prior to encoding; the zeros are not transmitted, but regenerated by the receiver prior to decoding.

TABLE 7.1   FEC Frame Delineation Symbols

Symbol	Description	Coding
/S_FEC/	Start of packet	/K28.5/D6.4/K28.5/D6.4/S/
/T_FEC_E/	End of packet, even alignment	/T/R/I/T/R/
/T_FEC_O/	End of packet, odd alignment	/T/R/R/I/T/R/

System Considerations and Reconciliation Sublayer

If the OLT were to be implemented as a PHY with a single MAC attached to it, this would cause serious trouble for any bridge to which the MAC were connected. Consider a MAC frame coming into the bridge from an ONU via an EPON port. The bridge associates the source address of the frame with the port on which it came in, that is, the EPON port. When, at a later time, another MAC frame comes in from an ONU, this time destined for the MAC address previously learned, it will *not* be transmitted back to the EPON port, because the bridge assumes that the frame was already received by all the stations on the attached "broadcast" LAN. As explained above, this is not the case; upstream transmissions are not received by other ONUs. Standard bridging has no way of accommodating an attached LAN that behaves as a broadcast LAN in downstream and as a point-to-point LAN in upstream.

The solution to this problem was designed in close cooperation with Working Group 802.1. Instead of one single MAC, the OLT would have a different dedicated virtual MAC for every ONU attached to the EPON, as shown in Figure 7.2. As a result, higher layers can consider the EPON as a collection of logical point-to-point links.[5]

From the individual MACs down to the OLT PHY, the logical point-to-point links share a common GMII; hence, a way to identify data frames for/from the different ONUs is required. The Logical Link iden-

---

[5] I think of an EPON as a formally conducted meeting: When the chair (OLT) speaks, everyone can hear it, even when she addresses just one member of the assembly. When a member speaks, he has to address the chair, even when the message is intended for an other member [35].

LLC or other MAC Client	LLC or other MAC Client		LLC or other MAC Client
MAC Control			
MAC 1	MAC 2	----------------------	MAC n
Reconciliation Sublayer			
	GMII		
PCS			
PMA			
PMD			

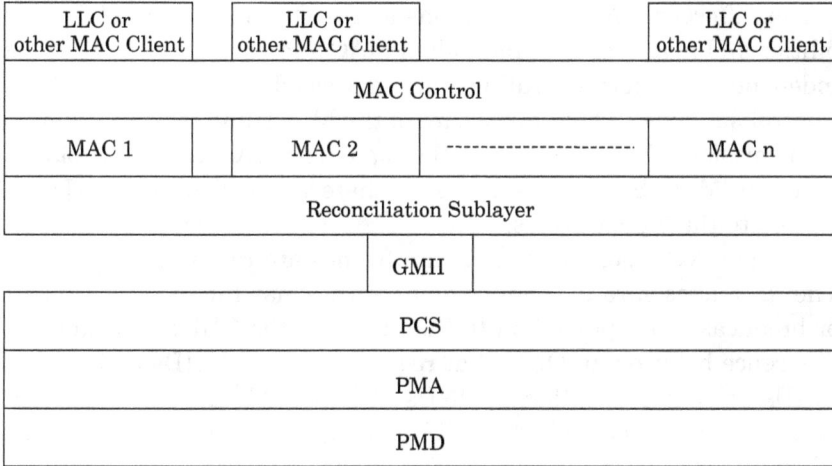

Figure 7.2 Architectural model of the EPON OLT.

tifier (LLID) was created for this purpose. Because the existing fields in the MAC frame could not be touched, part of the preamble had to be used to carry the LLID across the link:

- The first two octets of the preamble remain as they were.
- The third octet of the preamble is replaced by the start of LLID delimiter (SLD) field (0x5D).
- The fourth and fifth octet of the preamble remain as they were.
- The single-bit mode field and the 15-bit LLID are transported by the sixth and seventh octet of the preamble.
- The Start Frame Delimeter (SFD) octet is replaced by an 8-bit CRC calculated over octets 4 through 7 above.

This change to the preamble is applied by a newly specified Reconciliation Sublayer (RS) for EPON. Each of the logical MACs at the OLT receives only those frames that have an LLID that corresponds to that particular MAC. When a MAC transmits a frame, the RS adds an LLID to it to identify the MAC that it originated from. The RS thus ensures that there is a point-to-point tunnel between a single MAC Service interface at the OLT and a MAC Service interface at the ONU. In addition to allowing higher layers to access the EPON as a collection of point-to-point links, the modified preamble allows broadcasting of frames to all (or all but one) ONUs (see the section "Broadcast and LAN Emulation in PONs" below).

The different MAC instantiations at the OLT have full MAC capabilities and may even own individual MAC addresses, but they are not independent. Their operations are governed by a common MAC Control sublayer[6] which takes care of the MPCP and coordinates data transmissions. Nevertheless, at the top of the MAC Control sublayer, as many MAC Service instances as there are registered ONUs are offered to the higher layers.

At the ONU side, all downstream frames are received by the PHY. The RS makes sure that only frames with a matching LLID—unicast or broadcast—are passed on to the MAC. As the OLT cannot tell the difference between an ONU that registers several LLIDs and a set of LLIDs belonging to different ONUs, it is possible in principle to use different LLIDs for a single ONU. This approach may have benefits when different traffic classes must be polled on a regular basis [27].

This architecture, pure as it may be, poses a practical problem: How does one build an "EPON component" that contains all OLT functions from the MAC Service access point down to the fiber connector? Specifically, what interfaces must be provided on such a component to offer a large and variable number of MAC Service instances? Consider the following options.

1. Implement independent physical interfaces to the transmit and receive paths of as many MAC instances as there may be registered ONUs.

2. Implement a shared physical interface consisting of a data bus and an address bus in transmit and receive direction. The values on the address bus map to the LLIDs corresponding to the individual (virtual) MAC instances.

3. Connect each of the individual virtual MAC instances internally to a separate two-port MAC Relay, the other port of which is exposed to the outside world by means of a GMII.

4. Connect the variable number of individual virtual MAC instances internally to a single MAC Relay, and expose an additional connected MAC to the outside world by means of a GMII or XGMII.

5. Connect each of the individual virtual MAC instances internally to a separate IP entity.[7]

---

[6] The MAC Control sublayer is mandatory for 1000BASE-PX10 and 1000BASE-PX20. It is optional for other full-duplex systems, where it may be used for flow control (PAUSE frames).

[7] This layer-3 approach was proposed to the EFM Task Force [41].

Solution 1 is a direct implementation of the architectural model of the EPON OLT (Figure 7.2). It is, however, unpractical given the number of pins or connectors that would be required to wire 16 full-duplex MACs. The bus solution 2 solves that problem, but the lack of an appropriate standard bus that would guarantee interoperability with a bridging component makes this choice somewhat uncertain.

Solution 3 is cute, but it is wasteful and inefficient if the different GMIIs are going to be hooked up to the same bridge. Finally, solution 4 (illustrated in Figure 7.3) is not your typical PHY component, but it is the exact setup for which the multiple-MAC architecture was originally designed. As EPON OLT components are expected to show up in subscriber access multiplexers (CO equipment), interconnection to a bridge or some other kind of switching fabric will almost always be required.

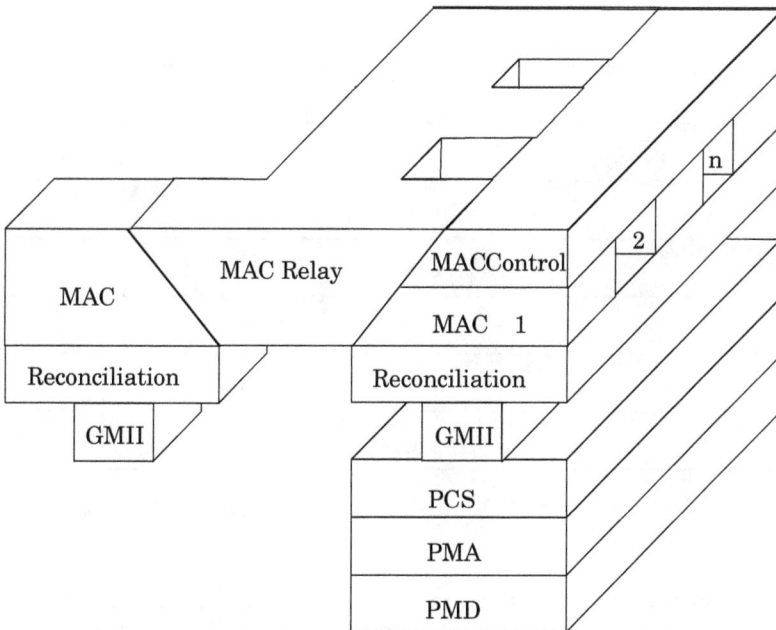

Figure 7.3   Connecting the EPON OLT to a MAC bridge.

Solution 5 is a solution in which the OLT is used at an end station (as opposed to a bridge). A single host may indeed connect to different IP networks, that is, it has a different IP address on its different layer-2 interfaces. Such a system is said to be *multi-homed*. If the multi-homed system has an EPON OLT Network Interface Card (NIC), it can be connected to up to 16 different IP networks via the same physical interface. For a personal computer, this may seem to be overkill, but a *router* is exactly the kind of network element that needs layer-2 connectivity with several separate networks to provide layer-3 connectivity between them. The EPON router, shown in Figure 7.4, is thus the layer-3 equivalent to the EPON bridge (solution 4), with the same close match to the theoretical architecture of the EPON. Routers may be used as subscriber access multiplexers as an alternative to bridges, if the additional cost (due to the complexity of routing protocols and forwarding algorithms) can be justified by the additional services offered to the subscribers.

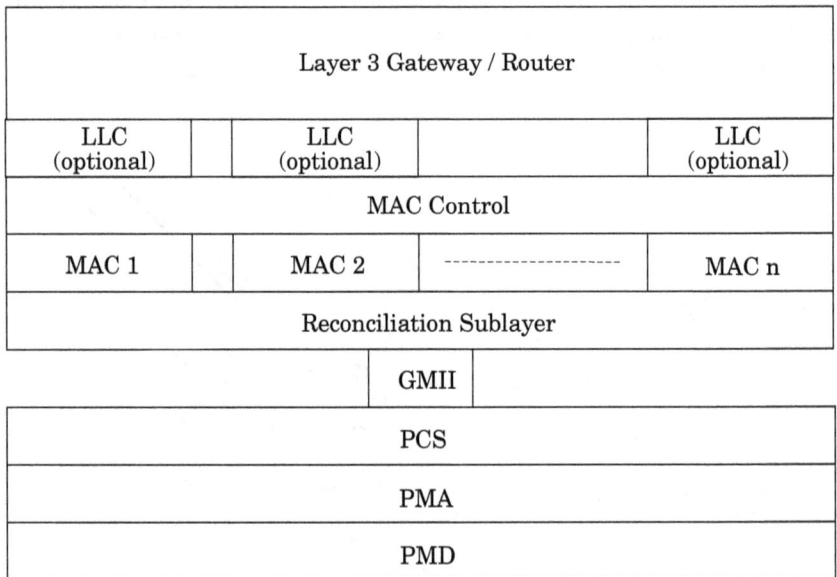

Figure 7.4   EPON OLT -based router.

As the reader can see, there is a fundamental difference between Figure 7.3 and Figure 7.4—this is not just a matter of bringing some variation into the artwork! The MAC Relay entity, which connects the different MACs in a MAC bridge, does not use the MAC Service to transmit or receive frames through a particular MAC. Instead, it uses the Internal Service Sublayer (ISS), the address-agnostic backdoor into the MAC. This is important because the MAC Relay needs to receive all frames that come in through a particular MAC, regardless of their MAC address—the frame filtering mechanism of the MAC Service is therefore undesired. Conversely, the MAC Relay needs to be able to send frames with any Source Address through the attached MACs—the Source Address completion of the MAC is therefore also undesired.

The result is a system that completely bypasses the MAC Service Access Points (which are still present and may optionally be used to access a port management entity). The bridge is a pure layer-2 solution. The router, on the other hand, uses the MAC Service to access its different "interfaces." Forwarding decisions will be based on layer-3 information, and all layer-2 functions will be left to the MAC.

Multi-Point MAC Control

Shared Networks versus Divided Networks

In an EPON, several subscribers use the same physical infrastructure to access the same OLT. It is therefore tempting to view EPONs as a kind of "shared networks," which Ethernet is historically very well suited to deal with.[8] Sharing, however, requires that all participants are mutually sensitive to each other's needs. This may be the case in an actual LAN, where Metcalfe's Law[9] holds, because the value of the network lies in the fact that *all* participants are and remain connected. Metcalfe's Law breaks down as soon as the connected users compete with each other for bandwidth. As an extreme example, consider a single user consuming all the bandwidth on a shared LAN; he would essentially be unable to interact with any other user, thus reducing the value of the network to its minimum.

Although the Internet as a whole is often considered to be a Metcalfe network, the subscriber access network by itself certainly isn't one. From the point of view of a (selfish) subscriber, the only thing that matters is being connected to the service he has subscribed to: a router providing Internet connectivity, a video-on-demand server, a digital television head-end, etc. Any other subscribers connected to the same access network are merely competitors for bandwidth. One must not expect a group of subscribers to be sensitive to each other's needs, and to voluntarily share the access medium in a fair way. To ensure fairness and a predictable level of service, a cen-

---

[8] In fact, this is even reflected in the name *Ethernet*, where the "Ether"—although it is now a deprecated physical concept—symbolizes a medium to which all participants have equal transmit/receive access.

[9] Metcalfe's Law, named after one of the inventors of Ethernet, states that the value of a network equals the square of the number of hosts connected to it.

tralized access control mechanism is needed. For EPONs, this centralized mechanism is provided by the Multi-Point Control Protocol (MPCP), by means of "GATE" messages (downstream) and "REPORT" messages (upstream).

Multi-Point Control Protocol

The MPCP was added to the MAC Control sublayer to allow ONU discovery, ranging, and coordination of upstream transmissions. A set of "REGISTER" messages is defined to allow new ONUs to join the PON and receive transmission grants. The OLT uses "GATE" messages to indicate to each of the ONUs when they are allowed to transmit. ONUs respond by sending "REPORT" messages (within their allocated burst slot), in which they indicate bandwidth requirements per priority queue (used for dynamic bandwidth allocation).

Just like the "PAUSE" frame mechanism, the MPCP protocol operates in the MAC Control sublayer, which resides between the MAC sublayer and the higher layers (such as the optional OAM sublayer and the MAC Control Client). The MAC Control sublayer is responsible for multiplexing the control frames into the data stream. At the OLT, a single MAC Control instance will connect to multiple MAC instances while offering the same number of MAC Service access points. In this manner, the MAC Control sublayer can coordinate the transmissions/receptions of all the associated MACs.

TABLE 7.2   MPCPDU Frame Structure

Field	Length	Description
Modified Preamble	8 bytes	See the section "System Considerations and Reconciliation Sublayer" and SFD
Destination Address	6 bytes	MAC Control multicast address or individual MAC address associated with targeted port
Source Address	6 bytes	Source address of the MAC from which the MPCPDU originates
Length/Type	2 bytes	Mac Control type = 0x8808
Opcode	2 bytes	See Table 7.3
Timestamp	4 bytes	See the section "Ranging and Timing"'
Data/Reserved/Pad	40 bytes	Opcode-dependent Field
FCS	4 bytes	32-bit CRC as specified for IEEE 802.3 MAC frames

When a frame is received from an underlying MAC, it is inspected as to whether it is MAC Control frame (type field = 0x8808) or a data frame. MAC Control frames are forwarded to the appropriate function; data frames are forwarded to the MAC Client corresponding to the MAC from which the frame was received. When a MAC Client transmits a frame over the MAC Control Service interface, the frame is forwarded to the appropriate MAC when this is allowed by the Multiplexing control block. Transmission of the data frame may be deferred until one or more MAC Control frames have been transmitted.

The five new MAC Control opcodes that have been specified for MPCP are shown in Table 7.3.

TABLE 7.3   MAC Control Opcodes for the Multi-Point Control Protocol

MPCP Function	Opcode
GATE	0x0002
REPORT	0x0003
REGISTER_REQ	0x0004
REGISTER	0x0005
REGISTER_ACK	0x0006

The GATE and REPORT functions are part of steady-state data transmission control. The REGISTER_REQ, REGISTER, and REGISTER_ACK functions are part of the discovery mechanism.

Discovery Process

The discovery process allows the EPON to dynamically update its logical topology when new ONUs are switched on or connected to the network. The OLT provides dedicated discovery timeslots during which newly attached ONUs can report their presence.

During such a discovery timeslot, which is announced by a broadcast GATE message with the discovery mode flag set to 1, several ONUs may simultaneously attempt to access the EPON by transmitting REGISTER_REQ messages. A random back-off algorithm is executed by all ONUs to reduce the probability of collisions.

Newly discovered ONUs are bound to a unique MAC at the OLT and are confirmed by means of a REGISTER message, sent over the broadcast channel to the requester's MAC address.[10] The REGISTER message contains the Logical Link ID and the synchronization time required by the OLT for that particular ONU.

After this confirmation, the ONU acknowledges registration by sending REGISTER_ACK during a regular transmit grant. From that moment on, regular data transmission can take place.

Multiplexing

Point-to-point emulation allows the different links of the EPON to behave as separate ports on a bridge at the OLT side. This is the preferred mode of operation in a commercial deployment in which the different subscribers have no desire to share the LAN. Downstream Multiplexing Control is used to make sure that only one MAC can access the RS, and thus the medium, at a time. The Multiplexing Control function inside the MAC Control sublayer provides "transmitEnable" signals to the different MAC Clients. Upstream Multiplexing Control is achieved by means of "GATE" and "REPORT" messages.

"GATE" messages are used to grant a certain ONU permission to transmit during a specified time window. The "GATE" message specifies the earliest time at which the ONU may turn its laser on (this is assumed to take up to 512 ns), and the time by which the laser shall be off (this is also assumed to take up 512 ns). An overlap may exist between "GATE" messages for the time required to turn the laser of the next ONU on. "GATE" messages are furthermore issued regularly to keep the watchdog timer awake.

"REPORT" messages are sent by the ONU to request bandwidth for its different transmit queues, or to simply let the OLT know it's still there if no request has been sent for the last 50 ms (to prevent it from being deregistered). The different transmit queues correspond to the different priority-bit settings used in priority-tagged or VLAN-tagged frames generated by IEEE Std. 802.1Q-compliant bridges or VLAN-enabled clients.

Broadcast and LAN Emulation in PONs

The special physical topology of an EPON, with its inherent downstream broadcast, can be used to advantageously emulate certain use-

---

[10] Prior to establishing the logical link, the ONU's MAC address must be used to indicate the destination of the MPCPDU. This was clarified in the first maintenance ballot of the EFM standard.

ful topologies at the service level. On the other hand, correct operation of MAC bridges, as specified in IEEE Std. 802.1D, requires certain extra precautions to be taken, in order to hide this special physical topology from the higher layers.

A Single-Copy Broadcast (SCB) MAC is built into the OLT; in addition to the MACs connected to each of the ONU MACs, the OLT has one additional MAC, which can be used for the purpose of broadcasting. The SCB MAC is used only in downstream; its datagrams are received by all ONUs. This is achieved physically by setting the mode bit in the EPON preamble (see the section above) to logical 1, and asserting the special LLID 0x7F (all ones).

Shared LANs, that is, LANs in which multiple stations are connected to the same physical medium (e.g., coaxial cable for 10BASE5 and 10BASE2) or to a set of interconnected repeaters and/or hubs, have as a fundamental characteristic that any frame transmitted by a station can be "heard" by every other station on the LAN (point-to-point links are a special case, in that they consist of only two stations). Bridged LANs, on the other hand, are partitioned in disjoint address spaces or collision domains (each of which may in turn be a shared LAN, a point-to-point link, or a bridged LAN) by virtue of the MAC address learning capability[11] in each bridge. By default, an EPON is none of the above.

The presence of the SCB MAC in the standard is an implicit specification of an EPON-based hub. Shared LAN emulation is a feature that must be implemented in higher layers; by providing appropriate filtering and forwarding rules at the level above the MAC, and using the SCB mechanism, a (relatively efficient) shared LAN can be emulated between all ONUs. For frames originating from an ONU, shared LAN emulation relies on the special "all-but-one broadcast" (or "PONcast") provided by the EPON Reconciliation Sublayer. By setting the mode bit in the EPON preamble to logical 1, and asserting an assigned LLID, all ONUs except the one addressed will receive the frame. For frames originating from outside the EPON, shared LAN emulation uses the true SCB mechanism (mode bit set to logical 1, LLID set to 0x7FFF).

Simply hooking up the SCB MAC to the bridge without any special precautions will certainly confuse the bridge and higher-layer protocols.[12] Although bridges routinely broadcast frames to all forwarding

---

[11] Source addresses of incoming frames are "learned", i.e. associated with the port on which they came in. Subsequent frames with a destination address that has already been learned, are forwarded to the associated port only, instead of being flooded to all ports.

[12] The MAC Service specification does not support a point-to-multipoint MAC.

ports, either for the purpose of configuring the network topology or at
the request of a higher-layer autoconfiguration protocol, there is no
standard way for a bridge to take advantage of the SCB mechanism.
Bridges see the EPON OLT as a set of individual MAC Service access
points—nothing less, and nothing more. When a bridge needs to broadcast a frame, it will copy the frame to each of these MAC Service access
points; in the case of an EPON, the frame will be transmitted to every
ONU separately. Even if the SCB MAC were exposed to the MAC Relay
in the bridge, there is no requirement in either IEEE Std. 802.3ah or in
IEEE Std. 802.1D that a layer-2 broadcast shall be translated into a
single MA-UNITDATA request toward the SCB MAC.

A clever EPON-bridge implementer will of course use the SCB mechanism for that purpose of rationalizing bridge-originated broadcasts
anyway. Although this approach introduces a point of noncompliance
in the design, this remains invisible to the other ports attached to the
bridge's MAC Relay function and to the ONU's MAC Service interface.
Hence, there will be no interoperability issues whatsoever when the
SCB mechanism is used for the intended application.

In addition to the SCB MAC, multicast MACs may be defined for
specific downstream channels. Again, this function has no companion
(yet) in the bridge specifications.

Ranging and Timing

The OLT hosts the central 32-bit clock, which is incremented after
every 16-bit transmission. Each MPCPDU transmitted by the OLT
carries a timestamp field equal to the value of the clock at the time of
transmission. At the ONU, the timestamps on received MPCPDUs are
used to set the ONU clock, which is also a 32-bit counter. Each ONU
also places a timestamp in every MPCPDU it transmits, which allows
the OLT to assess the round-trip time (RTT) of the link.

The MAC Control Client is notified of this RT, and may use this
information for the ranging process, which consists of measuring the
temporal distance of each of the ONUs. This is essential for optimization of the timing of upstream transmission slots. Ranging is performed continuously to allow compensating for RTT variation.

Dynamic Bandwidth Allocation

The way EPON is specified, the entity above the MAC Control sublayer in the OLT is in charge of scheduling downstream traffic and

deciding when each ONU is allowed to transmit upstream traffic. This function is called bandwidth allocation. Two important characteristics determine the quality of service (QoS) of the link between the OLT and any ONU:

- The fraction of the PON's total bandwidth that is made available to that ONU, i.e., the duration of the transmit windows granted in GATE messages (upstream) and the data transmitted in downstream, averaged over a longer period of time
- The average time between granted transmit windows, or between downstream transmissions (cycle time)

The bandwidth fraction determines the data throughput of the link (may be different in each direction), while the cycle time impacts the latency and jitter properties of the link (especially important for real-time applications). Neither the scheduling algorithm for downstream transmission, nor the way to allocate grants is specified in the EFM standard. It is up to the MAC Control Client to find a clever way to meet the different users' throughput and latency needs.

In static bandwidth allocation schemes, fixed amounts of bandwidth are allocated to each ONU, and all ONUs are served in a round-robin way. Dynamic Bandwidth Allocation (DBA), on the other hand, uses the information conveyed in the REPORT messages to optimize the use of network resources.

With DBA, the cycle time can be adaptive, in order to allow ONUs with a greater reported backlog more bandwidth. In that case, the bandwidth per ONU must of course be limited to prevent any single ONU from monopolizing the bandwidth. Conversely, a minimum amount of bandwidth must be allocated to each ONU in order to sustain normal operation. The bandwidth allocation algorithm may be further enhanced to take into account the relative weight of the requests for each traffic class (as indicated in the different queue values of the REPORT messages from the ONUs), in order to provide certain Quality-of-Service guarantees.

Several articles about the state of the art of Dynamic Bandwidth Allocation for EPON can be found in the August 2004 issue of *IEEE Communications Magazine*.

EPON, GPON, or Switched Ethernet?

Both the EPON standard and the GPON standard offer a way to efficiently transport Ethernet packets at Gigabit speeds. Components implementing each of these standards are hitting the market at approximately the same time (the 2004–2005 time frame), with a small lead for EPON. Both have to compete with the existing solution, consisting of switch-terminating point-to-point Ethernet—EFM or legacy—links.

An important question for system vendors, network integrators, and operators, is whether to use EPON, GPON, or switched Ethernet (sometimes called ESON, for "Ethernet Switched Optical Network"). To answer this question, one must take into account the subtle differences in features and possibilities inherent to each of these technologies, as well as the economics, which will partly be determined by the take-up of each technology in the following years. I will leave the fortune telling to the experts and focus on the technical facts.

Efficiency. Gigabit Ethernet PHYs (EPON and point-to-point) use 8B/10B coding, which ensures robust clock and data recovery at the receiver, at the expense of using up 20 percent of the line rate. Furthermore, each Ethernet frame is transmitted with a preamble and SFD (64 bits, which are of course reused to signal the LLID), and an inter-packet gap (96 bits), which add more overhead. GPON, on the other hand, uses the GPON Encapsulation Method (GEM), which adds only a 40-bit overhead to each Ethernet frame.

Burst mode. BPON and GPON use very short burst overheads (12 byte times at 1.244 Gbps, or 77 ns), while EPON requires up to 848 ns to turn the laser on and allow sufficient stable time for the ONU to lock its clock. Furthermore, GPON specifies a power leveling mechanism (equivalent to the upstream power back-off mechanism in 10PASS-TS), to reduce the required dynamic range at the OLT receiver. Here, GPON has clearly chosen high performance, whereas EPON has chosen low cost.

Flexibility. GPON clearly wins from EPON on flexibility; GPON has three optical power classes (20 dB, 25 dB, or 30 dB maximum optical loss), four different bit rates (155 Mbps, 622 Mbps, 1.244 Gbps, and 2.488 Gbps), and two TPS-TC options (GEM and ATM), while optical Ethernet PHYs have only a single bit rate and a single TPS-TC. The two EPON types, 1000BASE-PX10 and 1000BASE-PX20, roughly correspond to GPON Class A (20 dB maximum optical loss) and Class B (25 dB maximum optical loss).

Security. The GPON standard specifies an embedded encryption mechanism (AES). EPON does not address encryption at the PHY level, but MAC-level security is being addressed in a dedicated IEEE 802.1 Task Group (see Chapter 9 for details).

Cost considerations. EPON has deliberately borrowed as much as possible from the existing Gigabit Ethernet specifications, to allow for economical development of EPON components and a short time to market. The similarity of certain optical parameters of EPON and GPON components should also create some economic synergy between those two markets. In that respect, the success of each of the three high-speed packet network options could benefit the two other ones.

If EPON and GPON are difficult to tell apart, there seem to be obvious differences between PONs and switched networks. Nevertheless, it is not easy to come to clear guidelines for selecting PON or switched Ethernet in particular situations.

Both PONs and switched networks are deployed in tree topologies, the last "node" before the ONU being a passive splitter in the former case, and an active switch in the latter case. The benefits of the passive splitter are the fact that it does not require electrical powering (hence, no need for electrical cabling and battery back-up) and the lower maintenance requirements (which means lower operational expenditure per subscriber). These two factors make splitters suited for deployment at a long distance from the CO, possibly in an underground enclosure or on a pole. It is therefore commonly understood that PONs are suited for sparsely populated areas. If the same set of subscribers—with no electrical power available between the CO and the ONUs—were to be served by a switched Ethernet network, individual fibers or fiber pairs would have to be provided from the CO to each of the customers (increased capital expenditure), and there would be different sets of electro-optics for each subscriber at the CO (increased operational and capital expenditure).

If, on the other hand, electrical powering is available in the field, it is possible to deploy active equipment such as Ethernet switches in street cabinets. Similarly, if many customers are located in a single building (multi-dwelling unit, multi-tenant unit), the active equipment can be installed in a common indoor space, such as a basement. Furthermore, if the distance from the OLT to the customers is short, the extra cost of pulling individual fibers or fiber pairs to customers is small. The expenditure related to maintaining larger numbers of optical ports remains, but it is generally believed (and promoted by optical component vendors) that there is a case for the "active star" in MTUs/MDUs, cities, and other densely populated areas of development.

The ONU does not have to be located at the customer premises in either case. Copper technologies such as 10PASS-TS, VDSL, and ADSL2+ can bridge the gap between the last active point in the field and the customer's home or office. This is believed to be the economically feasible stepping stone between the "slower" xDSL flavors, typically served from the CO, and the fiber to the home solutions. The high-bandwidth xDSL links can be deployed from a cabinet that is fed by either a switched optical network or a PON. However, as the xDSL termination equipment is *active* equipment, it must be deployed from a powered cabinet, the presence of which suggests that a switched optical network would also be feasible. For MTUs/MDUs, the optical distribution can be completely passive, as the active xDSL equipment can again be placed indoors.

8

Management and OAM

*"For Ethernet service providers, managing
Ethernet services efficiently is key to a
sustainable business model."*
        —DIRCEU CAVENDISH [11]

Management Information Bases for EFM PHYs

Overview

A Management Information Base (MIB) of a system, as specified in a standard, is a structured list of the system's abstract resources and the attributes, operations, notifications, and behavior associated with these resources. Management of the system is assumed to be achieved through interaction of the manager with the objects in the MIB.

The actual managed resources of an implementation corresponding to these MIB objects may be very diverse. Registers in a chip, memory contents, the "state" of a process, physical presence or absence of a connected system, and many other things may all be represented towards the Manager as MIB objects. Hence, the MIB is merely a "view" of a collection of manageable resources. The translation of the implementation-specific characteristics of the system's components into standardized MIB object behavior is performed by an entity called the "Management Agent"; it interacts with the Manager through a management protocol, and with the managed resources through any appropriate means.

Although IEEE Std. 802.3-2002 does not specify a management protocol, it does specify the managed objects that must be visible to a Manager, when present, in great detail (Clause 30 of IEEE Std. 802.3). These managed objects, along with their behavior, attributes, actions, and notifications, are grouped into optional and mandatory "packages," which are in turn grouped into "capabilities." A system that claims compliance with IEEE Std. 802.3 management must implement the capabilities associated with the chosen port type. Within these capabilities, it must support all mandatory packages and it may support the associated optional packages. In certain capabilities, there are also conditional packages, which need only be supported if a certain optional feature or feature group is implemented.

A managed Ethernet port has to implement the Data Terminal Equipment (DTE) capability to manage objects related to the operation of the relevant sublayers in the data link layer. It includes one *oMACEntity* and one or more *oPHYEntity* instances. The Medium Attachment Unit (MAU) capability must also be implemented, to manage objects related to the operation of the physical layer. It includes the *oMAU* managed object class. These capabilities have been part of the base standard for a long time.

To manage the new Ethernet in the First Mile (EFM) functions, the Task Force created a number of new packages and a new capability. The conditional *Multi-Point Control Protocol Package*, *Operation Administration Maintenance Package*, and *Optical Multi-Point Emulation Package*, and the optional *Optical Multi-Point Emulation Monitor Package* are in the DTE capability. The MAU capability is extended with the optional *PCS Code Error Monitor Package* and the conditional *Forward Error Correction (FEC) Package*. The *Forward Error Correction Package* is only intended to manage the optional FEC feature in the *optical* EFM PHYs. The Reed-Solomon encoder in the 10PASS-TS PHY must be managed through the PME capability.

The entirely new PME capability was created to manage the objects of the Physical Medium Entity (PME),[1] multiple instances of which can be aggregated into a single PHY. PME consists of a mandatory *Basic Package*, an optional *PME Aggregation Package*, and a mandatory *10P/2B Package*.

Managing Ethernet PHYs through SNMP. In real-world implementations and deployments, the managed objects of Ethernet systems are most

---

[1] The PME comprises the lower three sublayers of the EFM Copper PHY: TC-sublayer, PMA, and PMD.

often accessed through their Simple Network Management Protocol (SNMP) counterparts, defined in Internet Engineering Task Force (IETF) MIBs. An overall architecture for the management of MIB-II objects, used with the SNMP protocol, is described in RFC 2571. The Structure of Management Information (SMIv2) is specified in STD 58, which consists of RFCs 2578, 2579, and 2580. The purpose of this section is to offer the reader insight into the structure and the main elements of the MIBs used to manage EFM ports. There is no intent to provide a complete listing of these MIBs, as they are still under development, and they can be freely consulted via the IETF website (www.ieft.org).

Managed objects used for managing network interfaces are defined in the Interfaces Group MIB (IF-MIB) (RFC2863); the number of network interfaces in the managed system is represented by *ifNumber*, each individual network interface being addressed by a number *ifIndex* between 1 and *ifNumber*.[2] The IF-MIB is cleverly designed to accommodate upward and downward multiplexing of sublayers, as is required for systems that support Link Aggregation, PME Aggregation, and bonding. This is achieved by having two tables, *ifTable* and *ifStackTable*. The former contains an individual row for each "interface" (i.e., each instance of any sublayer present in the system), while the latter identifies the relations between sublayer instances (in terms of superior and subordinate sublayers) by means of pointers. An inverted version of the interface stack, *ifInvStackTable*, is also available to make access to the different interface relations more efficient.

Specific extensions for managing Ethernet Medium Attachment Units are defined in the Ethernet-like interfaces MIB (RFC 3635) and in the MAU MIB (MAU-MIB) (RFC 3636). At the time of writing, the MAU MIB is being revised: The document is being split into a dynamic Internet Assigned Numbers Authority (IANA)-maintained part (IANA-MAU-TC-MIB), containing object identities and textual conventions, which can easily be updated as new port types are developed) and a more stable IETF-maintained part. The IANA-maintained part is being updated with management objects for 10GBASE-CX4, EFM, and 10GBASE-LRM.

Note that there is no RFC with common managed objects for all EFM port types. Separate RFCs cover EFM Copper, EPON, and OAM. The point-to-point fiber PHYs can be managed through the existing MAU-MIB objects for 100BASE-X and 1000BASE-X PHYs.

---

[2] In systems that allow dynamic creation and deletion of (logical) interfaces, indices outside this range are possible.

OAM MIB.   The MIB for OAM defines objects to control the OAM functions specified in IEEE Std, 802.3ah-2004, which are discussed in the section "Operations, Administration, and Maintenance." Seven different MIB groups are used to manage the OAM capabilities defined in EFM. These groups include MIB objects for all the mandatory features (discovery, state machines, critical events) as well as the optional functional groups.

- The *dot3OamTable* group is used to manage the primary OAM objects of the interface. It includes a read-only bitmap of the optional OAM functions supported by this Ethernet interface (unidirectional mode, remote loobpack, event notifications, variable retrieval) and status information about the discovery process (see the section "Discovery and Information").
- The *dot3OamPeerTable* group contains information about the status of the link partner's OAM entity.
- The *dot3OamLoopbackTable* manages the loopback function, which is discussed in the section "Remote Loopback." It allows the manager to initiate and terminate remote loopback mode, or to force the local OAM entity to ignore incoming loopback requests.
- The *dot3OamStatsTable* group contains statistics on the OAM frames exchanged through the interface.
- The *dot3OamEventConfigTable* and *dot3OamEventLogTable* groups are used to manage the event notification capability, which is discussed in the section "Event Notification." Notifications for Errored Symbol Period events, Errored Frame events, Errored Frame Period events, and Errored Frame Seconds events can all be enabled and disabled independently.
- Ethernet OAM notifications, which generate SNMP "traps" or alerts to the management entity, are set up through the *dot3OamNotifs* object group. This allows the management entity to passively keep track of locally or remotely detected events (such as symbol errors or frame errors). These events are exchanged over the link by means of OAMPDUs and signaled to the management entity by means of these traps.

Point-to-point copper MIB.   Although the 10PASS-TS and 2BASE-TL PHYs are specified by means of references to pre-existing xDSL standards, which already have dedicated MIBs defined for them, the EFM Copper MIB was rebuilt from scratch. This was done to avoid having to introduce awkward work-arounds for the various philosophical differences between the EFM Copper PHYs and their xDSL predecessors.

If there is a need to include particular managed objects from VDSL-LINE-EXT-MCM-MIB or HDSL2-SHDSL-LINE-MIB groups (for 10PASS-TS or 2BASE-TL, respectively), this is allowed as an option. For example, objects to perform performance monitoring, not provided by the EFM Copper MIB, may be borrowed from the corresponding xDSL MIBs if the implementer so desires.

The EFM Copper MIB is constructed around two object groups: *efmCuPme* and *efmCuPort*. The former represents a set of EFM Copper PMEs (10PASS-TS or 2BASE-TL TC/PMA/PMD combinations), while the latter represents a set of EFM Copper PCS instances, which may each be connected to multiple PMEs.

In addition to the *ifStackTable* defined in the IF-MIB, the EFM Copper MIB uses an *efmCuAvailableStackTable* to support the PAF discovery scenario described in Chapter 3. This object stores pointers to all PMEs that can be connected to a particular PCS, as reflected in the *PME_Available_register* specified in Clause 45. It is up to the discovery process—which is also managed by the EFM Copper MIB—to verify which PMEs are actually aggregated at the remote side and to make the appropriate local connections by updating the information in the *ifStackTable* object accordingly.

The port type of each PME is identified by its *ifType* value; it is *shdsl(169)* for 2BASE-TL or *vdsl(97)* for 10PASS-TS. The *ifType* of the PCS must be *ethernetCsmacd(6)*. The *ifAdminStatus* (in the IF-MIB) is used to initialize an EFM Copper PME or PCS port (see Chapters 4 and 5 for details on the initialization procedure), while the *ifOperStatus* object and the new *efmCuPmeOperStatus* object are used to monitor the progress of initialization and verify port status.

Finally, the concept of operational profiles introduced by EFM Copper (see the section "Management for EFM Copper" in Chapter 3) is maintained in the MIB objects *efmCu10PConfProfileTable* and *efmCuPme2BConfProfileTable*. These profiles are used to store a complete configuration, which can be applied to multiple EFM Copper ports. A complete 10PASS-TS profile consists of a bandplan and PSD mask, a reference PSD for upstream power back-off, a set of radio band notches, and a bit rate pair. A complete 2BASE-TL profile consists of a bit rate, a power level, a region (Annex A or Annex B of ITU-T Recommendation G.991.2), and a constellation size (16TC-PAM or 32TC-PAM).

EPON MIB. The "EFM EPON MIB" module is an extension of the Ethernet-like MIB, the IF-MIB, and the MAU-MIB. The EFM EPON MIB consists of three MIB groups:

- **MPCP MIBs:** The *dot3MpcpTable* contains objects used for the configuration and status verification of the Multi-Point Control Protocol. The *dot3MpcpStatTable* contains statistics relevant to the operation of the Multi-Point Control Protocol.
- **OMPEmulation MIBs:** This group concerns the point-to-point emulation feature in Ethernet Passive Optical Networks (EPONs). As explained in Chapter 7, point-to-point emulation is achieved by assigning a Logical Link Identifier (LLID) to each frame traversing the EPON. The *dot3OmpEmulationTable* contains the configuration and status objects, while the *dot3OmpEmulationStatTable* keeps track of the relevant statistics.
- **MAU MIBs:** This group defines the managed objects of EPON interfaces at the MAU level; it should be considered as an extension of the MAU-MIB mentioned above. It consists of a *dot3EponMauTable*, hosting the managed parameters of the EPON physical layer, and a *dot3EponMauType* type definition.

Additionally, there is an "EPON devices MIB" module defined in the same document, which contains objects that can be used to manage a *device*—such as a bridge—containing one or more EPON Optical Line Terminal (OLT) interfaces. A remarkable entry in the devices MIB is the *eponDeviceRemoteMACAddressLLIDTable*, which contains a table mapping Optical Network Unit (ONU) MAC addresses to LLIDs. Note that it stores the address of the ONU itself, and not of the end stations that may be attached to other ports of the bridge of which the ONU is a part. The ONU addresses are learned from incoming Multi-Point Control Protocol (MPCP) messages.

In an EPON-based bridge, a similar table must be present to store the associations between the MAC addresses of the end stations attached to the ONUs and the LLID through which they can be reached. Where a normal bridge learns associations between MAC addresses and bridge "ports," the ports are now replaced by the identifiers of the virtual point-to-point links inside the EPON. The details of this table are an implementation choice, so there is no associated object in the EPON MIB. At the management level, the bridge and the EPON will be represented as separate entities.

Point-to-point fiber MIB. From a management point of view, there is no difference between the EFM point-to-point fiber PHYs and the pre-existing optical Fast Ethernet and Gigabit Ethernet PHYs. Therefore, they can be managed by means of the same MIBs. If OAM is implemented, the OAM MIB additionally applies.

Operations, Administration, and Maintenance

Overview

Another accomplishment of the EFM Task Force is the specification of an optional "Operations, Administration, and Maintenance" (OAM) sublayer for Ethernet, positioned between the Media Access Control (MAC) sublayer and the MAC Control Client. Mechanisms to provide OAM have been part of telephony and provider-oriented protocol suites (such as ATM, Frame Relay, and SDH/SONET) for a long time. Ethernet, being a LAN technology, has been able to do without OAM up till now, but its recent move into the provider space has created a need for Ethernet-specific OAM. The possibility to manage *services* running over an Ethernet network will be a key factor in any service provider's decision to migrate to Ethernet [11].

Within the scope of its Project Authorization Request (PAR), the EFM Task Force could and did only define OAM over a single link. With the context of a subscriber access network in mind, the OAM protocol specified in the EFM standard assigns different roles to the provider station ("active") and the subscriber station ("passive"). It is, however, possible to use EFM OAM outside this context on any Ethernet link, if desired.

OAM Sublayer

As OAM is a sublayer, it relies on the lower layer services (i.e., the MAC service) to communicate with a peer sublayer on the other side of the link. This means that OAM uses in-band MAC frames to transmit and receive information. Upwards, the OAM sublayer provides two interfaces: First, there is a new MAC Service interface, which takes the place of the MAC Service interface at the top of the MAC or MAC Control sublayer, which the MAC Client would interact with if the OAM sublayer were not present; second, there is an OAM Service interface, which allows the OAM sublayer to interact with an OAM Client. The OAM sublayer is responsible for multiplexing data originating from the MAC Client and OAM Protocol Data Units (OAMPDUs) onto the lower MAC Service interface. This will cause a limited reduction in the net bandwidth available to the MAC Client.

Just like the Link Aggregation Control Protocol (LACP), the OAM protocol is a "slow protocol," which makes it suited for implementation in either hardware or software. OAM Protocol Data Units (formatted as shown in Table 8.1 may be sent up to 10 times per second,[3] and the same OAMPDU may be sent several times to increase the likelihood of reception in high Bit Error Ratio (BER) situations. OAMPDUs make use of the MAC Control service interface, and may therefore be delayed due to incoming PAUSE frames, which provide flow control across the link. OAM provides mechanisms to monitor link operation and health, and to improve fault isolation. The OAM Sublayer can be used with any of the EFM PHYs as well as with any other point-to-point Ethernet PHY.

TABLE 8.1   Format of an OAMPDU (Destination Address, Source Address, Type, and FCS are fields belonging to the MAC frame)

Field	Length	Description
Destination Address	6 bytes	01-80-C2-00-00-02 (Destination Address for Slow Protocols)
Source Address	6 bytes	Unique MAC address of the transmitting station
Length/Type	2 bytes	88-09 (EtherType for Slow Protocols)
Subtype	1 byte	0x03 (OAM)
Flags Field	2 bytes	Bits 15–5 are reserved; bit 4: Remote Stable; bit 3: Local Stable; bit 2: Critical Event; bit 1: Dying Gasp; bit 0: Link Fault
Code Field	1 byte	See Table 8.2
Data/Pad Field	42–1496 bytes	Data pertaining to the specified OAM Code, followed by padding if necessary
FCS	4 bytes	32-bit CRC, as specified for IEEE 802.3 MAC frames

The choice for a frame-based OAM mechanism was heavily debated in the EFM Task Force and its OAM Sub Task Force. An alternative approach based on messages coded inside the preamble bytes of regular data frames was supported by part of the Sub Task Force for two reasons: First, preamble-coded OAM messages would not reduce the net bitrate available to the subscriber; second, preamble-coded OAM

---

[3] This is a change with respect to IEEE Std. 802.3-2002, in which slow protocols were limited to 5 frames per second.

messages could be inserted into the data stream more quickly, as they do not have to be placed in a frame queue. As a compromise, a baseline proposal was offered to the Task Force comprising both "OAM-in-frames" and "OAM-in-preamble" for different sets of messages, only the former being mandatory. The Task Force adopted only the frame-based part of the proposal, as described in this section.

By virtue of its objective, OAM includes Remote Failure Indication (RFI), Remote Loopback, and Link Monitoring. Additionally, vendor-specific extensions are possible by means of organization-specific OAMPDUs. Although it was originally part of the discussion on what was then called OAM&P, provisioning is no longer supported by OAM. As OAM sits within the scope of IEEE Std. 802.3, it only addresses single link segments; OAM frames are addressed to a reserved multicast address and are never forwarded by a bridge (see Chapter 10 for details on reserved addresses).

Link fault notifications, critical event notifications, and dying gasp are all part of the RFI category. OAMPDUs containing critical events may be sent continuously, and may even violate the 10 frames/second rule.

Remote Loopback can be enabled or disabled by a single loopback control command. Arbitrary data transmitted by the MAC Control Client on the near side of the link will be reflected by the OAM sub-layer on the far side of the link, allowing the MAC Control Client to assess the BER of the link (which will be close to the end-to-end BER when the link BER is better than 10^6).

Link monitoring allows for exchange of information PDUs, event notifications, and polling of variables in the IEEE 802.3 MIB. Variable polling only supports the "GET" function; hence, it is not a method to perform station management.

OAMPDUs are untagged MAC frames characterized by a *Type* field value of 0x8809 (slow protocols), a *Subtype* field value of 0x03 (OAM), and a *Code* field to identify the type OAMPDU. The different types of OAMPDUs are summarized in Table 8.2. All types of OAMPDU contain a *Flags* field used to signal discovery variables and critical event notifications.

The information in Information and Event OAMPDUs is grouped in records consisting of a one-byte *Type* field (identifying the kind of information), a one-byte *Length* field (indicating the length of the entire record—all three fields—in bytes), and a variable-length *Value* field (the actual information). These records are hence called TLVs in the EFM standard.

TABLE 8.2   OAMPDU Code Field Values

Code*	Name
0x00	Information
0x01	Event Notification
0x02	Variable Request
0x03	Variable Response
0x04	Loopback Control
0xFE	Organization Specific

*All other code values are reserved.

OAM Client

The OAM sublayer provides access to the OAM protocol through the primitives OAM_CTL.request, OAM_CTL.indication, OAMPDU.request, and OAMPDU.indication. The actual protocol exchanges are carried out by an OAM client that invokes the OAM sublayer's primitives according to the rules and the state machines described in the standard. It is the OAM client's responsibility to respond appropriately to the peer OAM client's messages and to keep track of the reported events.

In order to respond to the peer's "GET" commands to retrieve MIB variables, the local OAM Client must be capable of accessing the local managed objects through a local station management entity (STA). This STA is normally optional, required only to locally manage, configure, or provision (GET/SET) the local station.

Configuring or provisioning any nonlocal station—assuming that the remote station supports management—requires higher-layer management protocols, such as the SNMP, as explained above. These protocols can be used regardless of whether the station to be managed is on the other side of an EFM link or on the other side of a global network. The Ethernet list of managed objects (Clause 30) or the corresponding IETF MIBs may be used for this purpose.

Unidirectional Operation

Pre-EFM 100BASE-X and 1000BASE-X PHYs are not allowed to transmit data when they are unable to detect an active link with a peer PHY. It is clear how this restriction can seriously limit the usefulness of certain critical event notifications, as critical events have a nature of occurring when there is no (more) active link. It was therefore a logical step for EFM to lift the ban on unidirectional links, albeit solely for the purpose of transmitting OAMPDUs (these OAMPDUs will have the "Link Fault" flag set). Support for unidirectional transmission of OAMPDUs is an optional feature in OAM. Unidirectional data links remain unsupported.

The necessary changes were made to the 100BASE-X PCS and the 1000BASE-X PCS (see Chapter 6) to allow unidirectional operation. It is the responsibility of the OAM sublayer to filter data frames when the link is unidirectional.

The copper PHYs 2BASE-TL and 10PASS-TS do not support unidirectional operation at all, but these PHYs have their own out-of-band management channels, which allow notification of critical events with a unidirectional nature.

OAM Protocol

Discovery and information.   Because OAM is optional, a mechanism is provided that performs OAM capability discovery. Within the OAM specification, there is a core that must always be present (consisting of the service interfaces), and a set of different features, which are individual options: remote loopback, unidirectional operation, link event notification, variable retrieval, organization-specific OAMPDUs, organization specific events, and organization-specific Information TLVs. The draft distinguishes DTEs in active mode and DTEs in passive mode. In active mode, the DTE will initiate the OAM discovery process and may send any kind of OAMPDUs. In passive mode, the DTE will wait for the remote device to initiate the discovery process, and send no other frames than information PDUs, event notifications, or responses to requests from the active side.

OAM is only enabled after a successful discovery. Discovery is successful when both sides agree by means of Information OAMPDUs on the set of features that is supported. Information is coded in OAM Information TLVs using the type designations shown in Table 8.3.

The individual features are selected through the *OAM Configuration* field, a single byte that consists of 3 bits reserved for future use, followed by a bitmap representing *Variable Retrieval, Link Events, OAM Remote Loopback Support, Unidirectional Support,* and *Active Mode* (zero indicates passive mode).

Each side also transmits the *OAMPDU Configuration* field, indicating the maximum OAMPDU size that it is willing to receive. Both sides are bound to observe the smallest of the two indicated maximum OAMPDU sizes. In this way, the maximum bandwidth of the OAM "channel" can be limited to any value between 5.760 kbps (10 frames per second of 64 bytes each, plus preamble) and 122.080 kbps (10 frames per second of 1518 bytes each, plus preamble) in 80-bps increments.

TABLE 8.3   Event Type Field Values

Type*	Name
0x01	Local Information
0x02	Remote Information
0xFE	Organization-Specific Event TLV

*All other code values are reserved.

The discovery process loops through a number of states, each of which are signalled periodically (once per second) to the peer OAM sublayer. The state machine resets when the link is lost or when no OAMPDUs are received for 5 seconds. The process consists of the following steps:

1. Before establishment of the link, the DTEs send minimum-size information OAMPDUs with the Link Fault bit set and without any information TLVs. This step requires unidirectional operation to be supported by the PHY (see above).
2. When the link comes up, a DTE configured in passive mode initially waits for an Information OAMPDU with the Local Information TLV.
3. When the link comes up, a DTE configured in active mode initially sends Information OAMPDUs that only contain the Local Information TLV.
4. After receiving an Information OAMPDU with the Local Information TLV, the DTEs send Information OAMPDUs with both the Local and Remote Information TLVs. The *local stable* flag is set and the *evaluating* flag is reset when the local OAM sublayer is satisfied with both local and remote settings. Note that there is no way for an unsatisfied local OAM Client to suggest other settings to the remote OAM Client; withholding the *satisfied* state is not really a negotiation tool, but rather a no-go indication; the discovery process cannot continue.

5. When the local DTE receives confirmation that the remote DTE is satisfied with the OAM settings, it enters the state in which all negotiated OAM capabilities are activated.
6. State 1 is re-entered whenever the local OAM sublayer is no longer satisfied with the OAM settings.

Critical events. Three kinds of critical event notifications are supported by EFM: link fault, dying gasp, and "unspecified critical event." These are all communicated in the *Flags* field of any OAMPDU.

- Link fault is signalled when the receive path is broken. Obviously, this signal is useful only if unidirectional links are supported, which precludes its use on EFM Copper links.
- Dying gasp is asserted when a device detects a local condition that will cause an imminent failure from which it cannot recover, such as loss of power. This signal may be used by the modem on the other side of the link to initiate a more graceful shutdown or to redirect data to a redundant path, if such a path is provisioned. The standard does not specify how dying gasp is to be used in the case of an aggregated EFM Copper link, but it seems to the author that it should be asserted only when the last member of an aggregate is about to terminate normal operation.
- Any critical events beyond those specified in the standard can be signalled to the other end as an "unspecified critical event." The use of this feature depends on the capability of the modem to detect such critical events. With respect to aggregated EFM Copper links, the unspecified critical event can be used to signal the unexpected termination of one of the aggregated PMEs, which is *not* the last active one (see dying gasp).

The EFM standard does not provide any more specific definitions of these critical events, nor does it specify how the receiving system should react to critical event notifications. These issues are left to the discretion of the implementer. The critical events are therefore no more than general indications of faults that require further attention from the network operator.

Event notification. EFM OAM uses Event Notification OAMPDUs to keep track of four standardized types of noncritical events, as well as organization-specific events. Together, these event notification capabilities provide the "link monitoring" function specified in the EFM Task Force's OAM objective.

The Event Notification OAMPDUs are generated by the OAM sublayer upon request from the OAM client (using the OAMPDU.request primitive); to improve reliability, the OAM client may send several identical Event Notification OAMPDUs. Information about events is coded in OAM Event TLVs using the type designations shown in Table 8.4. The mentioned time windows are measured in 100 ms intervals.

TABLE 8.4   Event Codes

Type*	Name	Description
0x01	Errored Symbol Period Event	Number of errored symbols exceeds a preset threshold within a preset number of received symbols.
0x02	Errored Frame Event	Number of errored frames exceeds a preset threshold within a preset time window.
0x03	Errored Frame Period Event	Number of errored frames exceeds a preset threshold within a preset number of frames.
0x04	Errored Frame Seconds Summary Event	Number of errored frame seconds exceeds a preset threshold within a preset time window.
0xFE	Organization-Specific Event TLV	—

*All other code values are reserved.

Errored frames are data frames with a detectable transmission error, such as a violation of the minimum or maximum frame size, a byte alignment problem, or an invalid FCS. Note that although only errors in the MAC-generated FCS are reported, these may actually represent errors detected by the PHY. When one of the lower sublayers detects an error in the received data by its own means, an error signal is relayed upwards over the successive interfaces. When it reaches the Reconciliation Sublayer by means of the RX_ER signal, it is passed on to the MAC as a deliberately corrupted FCS. In the case of aggregated EFM Copper links, the PME Aggregation Function inside the PCS will indicate certain errors by passing a frame composed of 64 zeros up to the MII. Such deliberate "garbage frames" will also be counted as errored frames.

Errored symbols are PHY symbols that are in some way detected as faulty. Symbols are defined as "the smallest unit of data transmission on the medium" (IEEE Std. 802.3), which may be different for different coding systems. In the EFM port types, symbols are defined as follows:

- For 100BASE-X and 1000BASE-X PHYs, the symbols are binary (bits)
- For 2BASE-TL, the symbols are 16-level or 32-level PAM
- For 10PASS-TS, the symbols are DMT-symbols

EFM OAM does not explain *how* symbol errors can be detected or counted. In 10PASS-TS, one DMT-symbol corresponds with one PMA frame, such that symbol errors can be related with PMA frame errors, which can be detected using the error detection mechanisms built into the PMA frame. For PHYs other than 10PASS-TS, the symbol error rate must be deduced from the occurrence of codegroup errors (the *aPCSCodingViolation* managed object was added to the Clause 30 MIB for this purpose) or from the MAC frame error rate.

In EFM Copper, multiple Physical Layer Entities (PMEs) may be aggregated into a single PHY. In that case, the symbol errors of each of the PMEs must be added together to get to the number that is to be reported through the event notification mechanism.

Remote loopback.   If loopback is supported, a DTE in OAM active mode may send Loopback Control OAMPDUs to its peer. Loopback Control OAMPDUs contain a single 1-octet Remote Loopback Enable/Disable command. These commands must be acknowledged by the receiving OAM sublayer by means of an Information OAMPDU containing the updated state information. When loopback is enabled in a device, the OAM sublayer transmits all frames received from the MAC sublayer,[4] back to the MAC. The MAC Client does *not* receive a copy of the frame.

Remote loopback allows the local OAM Client to perform fault isolation; if frames transmitted by the local OAM Client are correctly received again, the MAC-to-MAC path is known to be operational. Note that while loopback mode is active, no data are accepted from the MAC Client.

---

[4] Alternatively, from the MAC Control sublayer, if present.

Variable retrieval. The variable retrieval mechanisms allow OAM clients to read remote MIB object attributes (variables). The MIB objects are defined in a protocol-independent way in Clause 30 of IEEE Std 802.3 (a fully GDMO-compliant definition can be found in Annexes 30A and 30B, and an SNMP-compliant MIB exists for Link Aggregation in Annex 30C). The specification according to the Guidelines for Definitions of Managed Objects (GDMO) implicates a globally unique registration number, within the management tree assigned to IEEE 802.3: *iso(1) member-body(2) us(840) ieee802dot3(10006) csmacdmgt(30)*. Within this tree, attributes have branch designation *attribute(7)* and a variable leaf designation *leaf(x)*.[5] The tree can be expanded further downwards to specify an individual Ethernet port of a managed system, its MAC, its PHY, and any of the PMEs making up the PHY (in the case of aggregated EFM Copper).

Variables can be requested from the peer OAM entity by indicating a branch and leaf number in a Variable Request OAMPDU. The Variable Response OAMPDU repeats the branch and leaf number, followed by a 1-octet *width* field and a *value* field. The variable request may reference in its descriptor a single attribute, a package (set of attributes), or an object (set of packages). If a package or an object is requested, the returned variable container will contain all the concerned attribute values. If one or more variables can not be retrieved for any reason, the peer shall respond with an error indication.

Most Ethernet MIB objects have a second life as SNMP MIB objects (this mapping is explicitly indicated in the RFCs with Ethernet-related MIB objects, as discussed above). SNMP MIB objects also have globally unique registration numbers, which are assigned by IANA under a different tree: *iso(1) org(3) dod(6) internet(1) mgmt(2) mib-2(1)*. SNMP provides instruction to GET or SET variables. A possible management architecture would consist of a first DTE configured in active mode, with an SNMP agent connecting to network management, and a second DTE configured in passive mode, with no SNMP capabilities. The first DTE could then act as a management proxy for the second one, translating SNMP GET messages for the second DTE into the appropriate Variable Request OAMPDUs.

An example of management architecture encompassing a Central Office (CO) transceiver and a Customer Premises Equipment (CPE) transceiver is shown in Figure 8.1.

---

[5] Assigned registration arcs can be found at www.ieee802.org/3/arcs/802dot3_reg_arcs.pdf.

To Centralized Management System

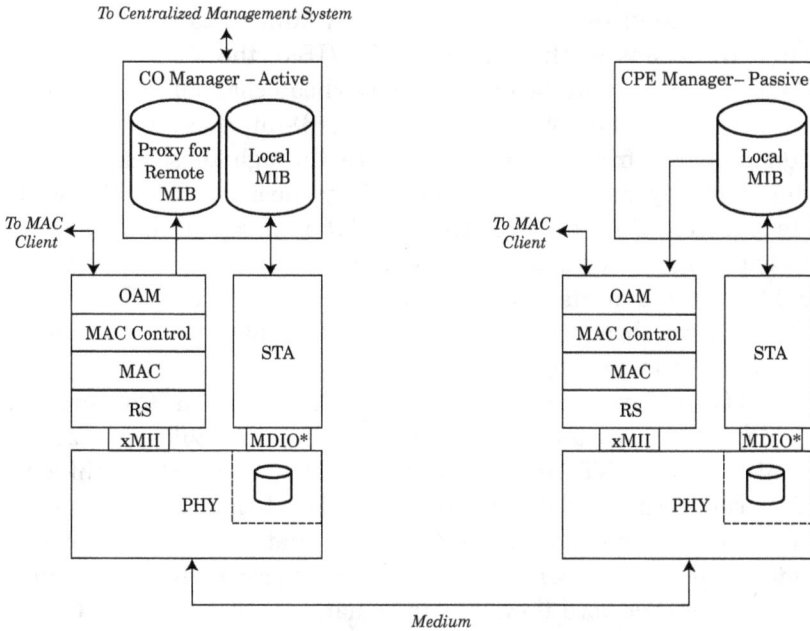

Figure 8.1   Example architecture for centralized management of CO and CPE equipment. The MDIO may be replaced by any other interface that gives access to the managed resources inside the PHY.

Applications of variable retrieval. The management architecture described above does not allow remote configuration of the passive DTE. However, management and OAM are optional and, by construction, are never required for normal operation. Other (mandatory) means of remote configuration are provided for the port types that need them.

In the case of EPON, configuration of the OLT is sufficient to determine the behavior of the ONU, as these PHYs and the MPCP are designed as a master/slave system in which the master determines the slave's behavior during normal operation.

In the case of EFM Copper, things are slightly more complex. The *configuration* phase is clearly CO driven. It is always the CO node that selects the operational parameters through the G.handshake exchange. In the case of 10PASS-TS, the 10PASS-TS-O device also calls the shots in the subsequent SOC-based activation. As for monitoring and maintenance during *normal operation*, every sublayer has its own peculiarities.

- The PMD and PMA were inherited from existing xDSL standards, from which they bring special ways to transport status information across the link: remote PMD and PMA failures can be communicated by means of the indicator bits (IBs), the VDSL Overhead Channel (VOC), and the embedded overhead channel (EOC), which are part of the data stream below the $\alpha(\beta)$-interface. The information gathered from the remote device through these channels is stored locally and may be accessed by means of the Clause 45 Management Data Input/Output (MDIO) registers, if implemented, as explained in Chapter 3. A very limited amount of PMD- and PMA-related information is reflected in the Clause 30 MIB and may be retrieved through the OAM entity, provided that it supports variable retrieval.

- The TC sublayer (see the section "Transmission Convergence Sublayer" in Chapter 3) was built on the same philosophy as the PTM-TC in ITU-T Recommendation G.993.1, which—unlike its ATM counterpart—did not have a dedicated channel for management information. The only information that the TC sublayer provides to its peer, is whether it is synchronized with the incoming 64/65-octet codeword flow; this information is stored in the Clause 45 MDIO registers as is the case for the PMD and PMA sublayers. The counters for coding violations and TC-CRC errors, on the other hand, are not communicated to the peer TC sublayer. This information can only be accessed by the device on the other end of the link, if it is made available to an OAM entity that supports variable retrieval.[6]

- The MACs on both ends of an EFM Copper link have no way of communicating with each other than through the data channel. This channel is only suitable for error notifications by means of deliberately corrupted FCS sequences, as explained earlier. Far-end MAC-related management information can again only be retrieved through that MAC's OAM entity, provided that it supports variable retrieval.

As a conclusion, the "optional" OAM sublayer seems to be a practical requirement for EFM Copper implementations. With an EFM Copper CPE that does not support variable retrieval, the network operator has no way of detecting errors that are specific to the TC sublayer without going to the customer's premises (a so-called "truck roll," which is a very expensive and hence undesirable solution) and connecting directly to the CPE's management interface.

---

[6] There is no object counting the TC-CRC errors in the original IEEE Std 802.3ah-2004, but one was added in 2005 in the 802.3REVam revision project.

Organization-specific OAMPDUs. The EFM OAM specification was conceived in such a way that it can easily accommodate vendor-specific extensions at different levels, for example, to exchange information that is not otherwise accessible. Vendors are identified by means of their Organization Unique Identifier (OUI), a 24-bit label, which is also used to identify the vendor in the MAC address of network equipment, as described in IEEE Std. 802-2001.

Information OAMPDUs and Event OAMPDUs may use Organization-Specific TLVs (indicating the vendor's OUI, of course) to indicate information or events that are not supported by the standard TLVs. In addition, there is an Organization-Specific OAMPDU, in which the first three bytes of the *Data* field contain the OUI and the rest of the field is at the discretion of the vendor.

End-to-End OAM

Overview of Ongoing Activities

For the management of a larger network, different provisions must be taken. Both ITU-T Study Group 13 and IEEE Working Group 802.1 have started activities to define end-to-end Ethernet OAM. Every effort is being made by the participants of both projects to keep the work aligned, but important differences in terminology and philosophy between the two Standards Development Organizations (SDOs) necessitate the development of two distinct documents. The IEEE 802.1 side of this activity is summarized in the section "Connectivity Fault Management."

IEEE projects P802.1AB and P802.1ag are specifying protocols that have a lot of similarities with OAM, as described in the previous section. Being developed inside the IEEE 802.1 Higher Layer Interfaces Working Group, these standards will take the network view, rather than the link view. By appropriately combining the information exchanged on individual links, the network operator will be able to detect and localize problems that occur anywhere in the network under her supervision. Where a virtual private LAN service (e-LAN) is being offered through a provider network, the subscriber is additionally capable of managing the end-to-end subscriber network independently of the provider's management operations.

Connectivity Discovery

The Link Layer Discovery Protocol (LLDP) is a one-way protocol that allows stations (including repeaters, bridges, access points, routers, telephones, and regular end stations) on a LAN to *advertise* their capabilities, or those of the system they are a part of, and the information required to manage that system. The messages defined in this protocol allow the receiving stations to gain a view of the topology of the network to which they are attached.

LLDP-enabled systems can transmit LLDP data units (LDPDUs) containing information about their own configuration. Transmission happens autonomously; there is no mechanism defined for a system to request configuration information from a remote system. Systems can operate in three modes: transmit only, receive only, or transmit and receive. For each port, the network manager decides which management objects associated with that port are to be transmitted through LLDP. Transmission occurs automatically upon expiry of a timer, or upon any change in the management object. Receiving stations can store information about other stations in their LLDP remote systems MIB.

The goal of the LLDP protocol is to enable network users to detect configuration inconsistencies that may degrade or prohibit certain network functions. Furthermore, users can obtain the information required to remove such inconsistencies with the help of higher-layer management protocols.

LLDPDU format.   LLDP uses TLV objects to encapsulate the information exchanged in LDPDUs. Note that the format of the TLVs used in LLDP is different from the format used in EFM OAM. TLVs used in LLDP have a 7-bit *TLV type* field and a 9-bit *TLV information string length* field, followed by a variable-length *TLV information string* field. The length field covers only the length of the *TLV information string* field.

Three TLV types are mandatory in every LDPDU (*Chassis ID*, *Port ID*, *Time-to-Live*), as they identify the source and the freshness of the transmitted information. The addition of other TLVs is optional. An *End-of-LDPDU* TLV must be used to pad LDPDUs that would otherwise be shorter than the required 46 bytes. Ethernet's intrinsic padding mechanism cannot be used here, because it would require the length before padding to be indicated in the MAC frame's *Length/Type* field, while LDPDUs are transmitted with a dedicated EtherType in the *Length/Type* field. The *End-of-LDPDU* TLV has the *TLV type* and *TLV information string length* equal to all-zeros.

The four mandatory TLVs listed above, along with five optional TLVs (*Port Description*, *System Name*, *System Description*, *System*

Capabilities, and *Management Address*), form the "basic management set," which must be supported by every LLDP-compliant system.

Other optional TLVs are specified in two organizationally specific sets: one for IEEE 802.1-related information and one for IEEE 802.3-related information (see below). Finally, organizationally specific TLVs beyond those listed in the standard may be defined by individual organizations. Support for the organizationally specific TLV format is mandatory.

Just like OAMPDUs, LDPDUs are constrained to a single link of the customer network—they use a link-constrained multicast address (see Table 8.5) from the range of reserved addresses defined in IEEE Std. 802.1D. The frames may nevertheless be forwarded by S-VLAN–aware bridges (provider bridges), which observe a different set of link-constrained multicast addresses. This implies that two customer bridges that are connected through a provider network may exchange LLDPDUs with each other as if they were on the same customer LAN. See the sections "MAC Bridges" and "Provider Bridges" in Chapter 10 for more information about link-constrained addresses.

TABLE 8.5   Format of an LLDPDU (Destination Address, Source Address, Type, and FCS are fields belonging to the MAC frame.)

Field	Length	Description
Destination Address	6 bytes	01-80-C2-00-00-0E (Destination Address for LLDP)
Source Address	6 bytes	Unique MAC address of the transmitting station
Length/Type	2 bytes	88-cc (LLDP Ethertype)
Chassis ID TLV	variable	Chassis identifier
Port ID TLV	variable	Port identifier
Time-to-Live TLV	4 bytes	Number of seconds that the recipient LLDP agent should consider information from this transmitter valid
Optional	variable	Optional TLVs and End-of-LLDPDU TLV if necessary
Pad Field	variable	Padding if necessary
FCS	4 bytes	32-bit CRC, as specified for IEEE 802.3 MAC frames

IEEE 802.1 organizationally specific TLVs. The IEEE 802.1 Organizationally Specific TLV set provides the semantics to advertise

information about the VLAN configuration of the system. As S-VLAN–aware bridges are transparent for LLDP, there is no need for TLVs containing provider bridge configuration information.

Port VLAN ID. The default VLAN ID that will be assigned to incoming untagged or priority-tagged frames (see the section "VLAN Bridges" in Chapter 10).

Port and Protocol VLAN ID. Defined only for bridges that support Port-and-Protocol-based VLAN classification. This TLV carries the PPVID number.

VLAN name. The VLAN name is a plain text name of up to 32 characters, with local significance only. The VLAN name is not part of a normal VLAN-tagged frame, so VLAN-aware bridges need not attach any special meaning to it.

Protocol identity. The protocol(s) that can be accessed through this port, identified by their Ethertype or LLC designation. This is very useful for troubleshooting, as protocol stack mismatches will cause situations where no communication is possible on a link.

IEEE 802.3 organizationally specific TLVs

The IEEE 802.3 Organizationally Specific TLV set provides the semantics to advertise information that is specific to Ethernet end stations, other than the information required to set up the link (which is exchanged through auto-negotiation).

MAC/PHY configuration/status. Indicates both the capabilities and the selected settings of the transmitting IEEE 802.3 node with respect to half- or full-duplex transmission and bit rate. It also indicates whether these settings were selected manually or through auto-negotiation.

Power via MDI. Ports of the types 10BASE-T, 100BASE-TX, and 1000BASE-T can support DTE powering via the Medium Dependent Interface, also known as "Power over Ethernet," as specified by the 802.3af amendment. This TLV allows the IEEE 802.3 node to advertise its Power over Ethernet capabilities to its peer. A number of bits in the TLV are reserved for future use.[7]

Link aggregation. This TLV expresses the link's capability to be aggregated, its current aggregation state, and the port identification of the

---

[7] A "Power over Ethernet Plus" study group is active at the time of writing.

aggregation, if applicable. Note that there is no corresponding TLV for PME Aggregation, as described in section "PME Aggregation for EFM Copper" in Chapter 3, as PME Aggregation occurs at lower layers and is hence invisible to the MAC Service interface.

Maximum frame size. A standard compliant Ethernet port must be capable of processing frames up to 1518 bytes (not VLAN enabled) or 1522 bytes (VLAN enabled). The use of this TLV to advertise values outside the IEEE 802.3 standard is permitted; this flexibility is provided to accommodate future expansions of the Ethernet frame. It could also be useful to advertise the capability to use so-called "jumbo frames": nonstandard frames larger than 1522 bytes, used to transport data more efficiently on high-speed inter-router links.

Effect of receipt. Upon receipt of an LDPDU, the LLDP entity runs a number of checks on the data to validate the received TLVs. Violations of the formating rules may cause individual TLVs or the entire LDPDU to be discarded. The information from validated TLVs is used to update the locally stored information concerning adjacent systems.

Connectivity Fault Management

Connectivity Fault Management (CFM) provides end-to-end management of faults, including detection, verification, isolation, and notification.[8] At the time of publication of this book, the P802.1ag project is still in a very early stage (editor's drafts); I will therefore limit the discussion of CFM to a high-level overview. The CFM standard will be an amendment to IEEE Std. 802.1Q *as amended by project 802.1ad*. This means that the operational context of the protocols and procedures defined by the CFM standard is that of *provider-bridged networks*.

The problems with end-to-end management in a multi-provider environment such as the one shown schematically in Figure 8.2, are that different providers have limited OAM access to each other's equipment, and that one person's end point is another person's intermediate hop. The virtual private line or LAN service experienced by the customer (assuming a business user for now), is provided by a service provider, whose service may very well depend on a chain of smaller regional networks owned by different operators. Each of these

---

[8] Recovery is covered by RSTP in IEEE 802.1 standards and Ethernet Fast Protection Switching in ITU-T.

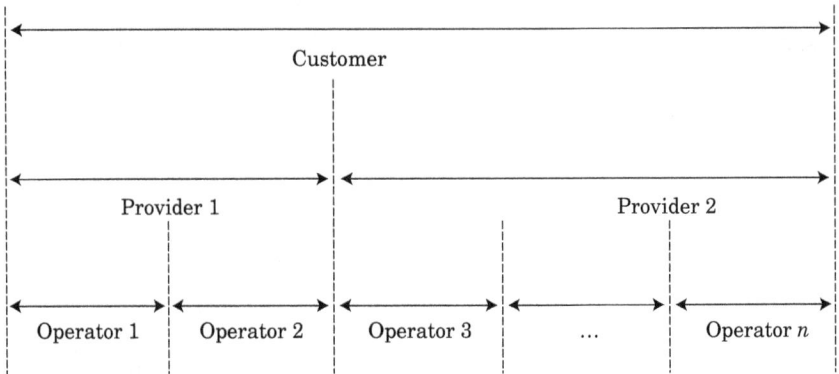

Figure 8.2   Different Maintenance Association levels in end-to-end OAM.

operators needs the capability to manage the equipment and connectivity in the network they control; the service provider needs the capability to manage the service spanning multiple operators' networks; the customer needs the capability to manage the connectivity between its various interconnected sites.

For end-to-end management to work, a "stack pointer" is required to indicate to participating nodes whether an end-to-end OAM flow is to be terminated at that node or at a subsequent node. To this purpose, the notion of Maintenance Association (MA) is introduced, which is understood to mean a connected part of the provider network (i.e., a network consisting of provider bridges, as defined in Chapter 10) in which the connectivity fault management protocol operates unobstructed. A Maintenance Association level is associated with each MA, such that higher-level MAs can contain lower-level MAs; overlap and containment of a higher-level MA within a lower-level domain is not allowed. A "domain" encompasses all maintenance entities under the control of the same administrative authority (e.g., a customer, an operator, a service provider).

CFM messages carry an indication of the MA level in the CFMPDU.[9] The boundary of a Maintenance Association is formed by a set of "Maintenance End Points" (MEPs), from which the CFM messages for that particular Maintenance Association originate, and where the CFM messages of that domain must be terminated. Maintenance Intermediate Points (MIPs) are managed provider bridges inside the Maintenance Association boundary. All Maintenance Points pass on as ordinary data any CFM messages with an MA level greater than their own. A given Maintenance Association may consist of two or more MEPs.

---

[9] In the current draft, this is a 3-bit field. Expansion of the field to 4 bits is under discussion.

The CFM function is a client of the MAC Service (or the Internal Service Sublayer) attached to a bridge port; hence, its messages are transported in MAC frames (as is the case for EFM OAM and LLDP). Multiple MAs, associated with different MA levels, may be attached to the same bridge port. CFM messages will carry a unique CFM Ethertype (value to be specified) and be addressed to the CM multicast address (for heartbeat messages, see below), the CM hop-by-hop address (for TraceRoute messages, see below), or the unicast address of a specific maintenance port. The last three bits in the CM multicast addresses replicate the MA level of the message, which facilitates immediate classification at the receiving port. Within an MA, the managed *services* are identified by their VLAN ID (VLANs are explained in more detail in Chapter 10). Just like EFM OAM and LLDP, CFM encapsulates the actual information in TLV records.

Fault detection.   Fault detection is accomplished through a continuity check mechanism. Every MEP of a MA periodically broadcasts a Continuity Check Message (CCM) or "heartbeat" to all other MEPs within the MA. The unexpected loss of heartbeat signals from a particular MEP or set of MEPs indicates a network fault in the MA. (MEPs that are taken out of service need to signal this to the rest of the MA, to avoid triggering a false fault detection.)

The CCM message may contain a *MEP List* TLV, specifying the next-level MEPs that are believed to become unreachable upon failure of the port to which the transmitting MEP is attached (this information is important when the MA has to decide whether to generate an Alarm Indication Signal for a given failure). The receiving MIP or MEP uses the information in CCM messages to populate a database of next-level MEPs that can be reached through the port on which the CCM was received. The database entries expire after the time specified in the *Lifetime* TLV.

Fault verification.   Fault verification is accomplished by a mechanism equivalent to the IP "ping" packet. The current draft of 802.1ag calls this the Loopback Message (LBM), which is unfortunate because of the possible confusion with the EFM OAM loopback mode, which is something entirely different. The LBM is a unicast message sent from a MEP to any other MEP or MIP. At the receiving maintenance point, the LBM is verified, and a Loopback Reply (LBR) is sent back to the source maintenance point. Readers familiar with IEEE Std. 802.2 will note that this loopback capability is very similar to the TEST and XID commands of the Logical Link Control (LLC) sublayer. These LLC commands are unpopular because they can be sent to group addresses, which means that improper use will easily lead to serious degradation or denial of service.

Linktrace. Another feature that is well-known from IP is the Linktrace Message (LTM) function. Unlike its IP counterpart "traceroute," the 802.1ag LTM is a single message, addressed to its final target destination, which is intercepted and processed at each hop before it is forwarded. This behavior is different from normal data forwarding behavior, which implies that some connectivity problems may not be accurately detected through Linktrace. Each intercepting bridge puts its own MAC address in the source field of the LTM frame, while retaining the original source address in the source TLV of the frame. The destination maintenance point terminates the LTM and sends back a Linktrace Reply (LTR) frame.

Alarm indication signals. Maintenance End Points can be configured to generate Alarm Indication Signals (AISs) upon detection of loss of connectivity. This is a powerful tool to distribute information about link failures to other network elements that support a common end-to-end service, regardless of whether the failure is permanent or temporary, and whether it can be automatically resolved by a spanning tree protocol. AIS also informs other (higher-level) Maintenance Points in the same network element, including the higher-level Maintenance Points attached to the same bridge port (in which case the AIS message is called a "virtual AIS"). The AIS expresses the fact that a lower-level Maintenance Association is taking care of the failure, thus avoiding the generation of redundant fault alarms for the same failure. For a given MA, loss of connectivity occurs when one MEP stops receiving Continuity Check Messages from the other MEP.

A few simple rules of link failure logic determine the requirements for an AIS protocol:

- The failure of a single physical link will have consequences for all the Maintenance Associations that rely on it, that is, the higher-level Maintenance Associations, but not for other individual physical links.
- The failure of a higher-level Maintenance Association will probably be caused by a lower-level failure, but it will not affect a neighboring Maintenance Association of the same level.

- As loss of connectivity will be detected at every affected Maintenance Entity level, information must be exchanged between the Maintenance Entities to identify the common cause of the failure and avoid a cascade of fault alarms.

When a loss of connectivity is detected by a MEP, AIS messages are transmitted periodically by that MEP at a predetermined rate, directed away from the direction of the lost Continuity Check Messages (which implies that the MEP at the other side of the failed link has to find out about the failure by its own means). The messages carry a TLV containing a list of the MEPs directly above the MA level of the detecting MEP, to which connection has been lost. This list is built from the MEP List TLVs in all CCM messages received from the MEP with which contact has been lost. It is important to distinguish between the MA level of the sender of the AIS message, as expressed in the header of the CFMPDU, and the MA level of the MEPs that are being declared out-of-reach due to the failure that is the object of the AIS message.

As the AIS message is of importance to the higher-level Maintenance Points that rely on the failed link for connectivity to their peers, it must be properly processed by all receiving MIPs and MEPs.

- As stated before, lower-level Maintenance Points pass higher-level messages on and/or up as if they were ordinary data; this is justified by the fact that a failure on one link does not affect the connectivity of Maintenance Associations that do not rely on it. Same-level MIPs must not take any particular action either, because connectivity within their Maintenance Association is guarded by the MEPs.
- A same-level MEP, on the other hand, may generate a fault alarm to alert the network administrator of a failure to receive Continuity Check messages. The alarm is sent periodically and for the duration of the fault.
- Next-level MEPs (i.e., MEPs whose MA level corresponds to the MA level of MEPs in the transmitted MEP List TLV) behave in the same way as same-level MEPs: they detect loss of continuity and generate an AIS toward the exterior of the MA (up in the levels) to suppress alarms generated in higher levels by the detected loss of continuity. AIS is always terminated by MEPs. It is regenerated if loss of continuity is detected.
- Higher-level MEPs must check whether the MEPs that are being declared "out-of-reach" in the AIS message are not accessible via a

new, different network path (they would know this, because they receive Continuity Check Messages from the lost MEP via the new path). MEPs that have become reachable via a new path are removed from the AIS message before it is passed up.

An AIS database entry will either expire (when it is not renewed before the time expressed in the *Lifetime* TLV), or be "revoked" by the transmitter when the failure is resolved, through transmission of a new AIS message with a zero lifetime.

Network OAM

With the current state of affairs, there are three mechanisms to gather information about the network elements around a certain point:

1. EFM OAM provides link monitoring, remote failure indication, remote loopback, and variable retrieval over *a single Ethernet link*
2. LLDP provides basic information on VLAN configuration, Ethernet configuration, and the management address of *adjacent ports*.

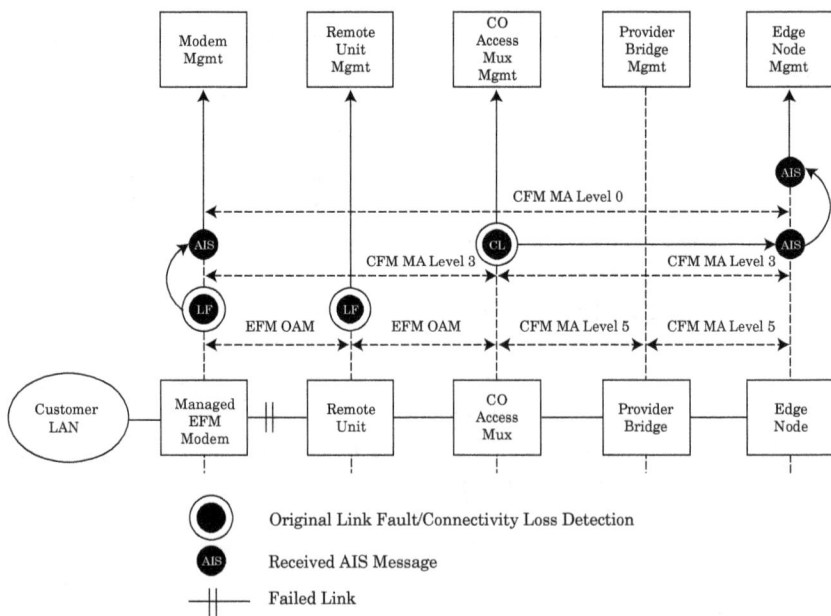

Figure 8.3 Coexistence of EFM OAM and CFM in a managed access network: message flows.

3. CFM provides fault detection and fault verification for the MIPs and MEPs *within the same management Maintenance Association.*

All three have to be combined to optimize the control the network operators have over the network. In a network in which some of the bridges do not have CFM capabilities, a straightforward combination of CFM and EFM OAM would be the use of link fault indications (EFM OAM) to generate outward AIS messages (CFM). IEEE 802.1ag defines an "AIS Convergence Function," which translates any available fault information—in particular from the EFM OAM entity, which can be considered as a very simple physical-layer MEP—into a meaningful AIS signal. In the case of a failure at the physical level, all higher-level maintenance entities must be considered out of reach.

Chapter

9

Security in
the First Mile

*"Encryption is the only way to protect our
privacy and guarantee the success of the
digital marketplace. The art of secret
communication, otherwise known as
cryptography, will provide the locks and keys
of the Information Age."*   —SIMON SINGH[1]

NOTE TO THE READER: At the time of writing, the MAC Security and Key
Security projects are ongoing. As a result, this chapter is partly based
on drafts, which are subject to change. I have therefore tried to focus
on concepts rather than details. Make sure to consult the most recent
draft or published standard for any recent changes.

Introduction

Overview of Security Issues

In subscriber access networks, it is not acceptable that frames
intended for one end station are visible to other end stations. To keep
this from happening, additional measures must be taken that are not
currently specified in IEEE Std. 802.1D or IEEE Std. 802.3. But eaves-
dropping is not the only security threat to the network; address spoofing, tampering with content, and any actions that may disrupt billing
and accounting are equally serious concerns.

---

[1] From the introduction of *The Code Book*, by Simon Singh, Fourth Estate, 1999.

In the most general terms, the goal of the IEEE 802 security standards is to guarantee that the parameters received in a MA-UNIT-DATA indication are identical to the ones transmitted in the corresponding MA-UNITDATA request by an authenticated user. Additionally, the VLAN standards (IEEE Std. 802.1Q and its upcoming amendment 802.1ad, discussed in more detail in Chapter 10) provide means to confine data traffic to certain portions of a physical infrastructure, called Virtual Bridged LANs (VLANs).

It may seem odd at first that the Ethernet in the First Mile (EFM) Task Force did not take on the task to provide security for subscribers on an Ethernet-based access network, but the strong belief in the layering paradigm that lives in the 802.3 community, led the Working Group to drop this topic from the Task Force objectives.[2] Security is traditionally a higher-layer issue, but as computing power becomes cheaper and more readily available, cryptographic functions are making their way down the protocol stack: from the application layer (e.g., encrypted email, custom crypto in end-user programs) down to the session layer (e.g., https, SSL), to the network and transport layers (e.g., IPSec), and in some cases further down to the data link layer (e.g., WEP).

In response to the security issues raised by the EPON specification, an Executive Committee Study Group (ECSG) was formed to investigate the need for an LMSC standard on "Link Security," work that was later picked up by the IEEE 802.1 "Higher Layer Interfaces" Working Group in the projects on MAC Security (P802.1AE) and Key Security (P802.1af, an amendment to IEEE Std. 802.1X "Port Based Access Control").

Encryption Basics

Encryption is the process of rendering a message unintelligible by means of a *cipher*, an algorithm with a certain set of parameters. Encryption is useless of course, unless the intended recipient has a way of decoding the original message, by using his knowledge of the cipher with which the message was encrypted.

The conventional cryptographic algorithms are "secret key algorithms," which use the same parameter (a cryptographic *key*) for both encryption and decryption. This requires the key to be a shared secret

---

[2] A fear of coming up with a security scheme that would be broken soon afterwards—as happened to the Wired Equivalent Privacy (WEP) scheme, invented by Working Group 802.11 for its wireless LANs—was another reason not to get into this issue. The remedy for this problem is to specify a flexible framework for security, allowing the use of different encryption algorithms according to the needs of the moment.

between the transmitter and the receiver of the message. An example is the Federal Data Encryption Standard (DES), which uses a 56-bit key. The weakness of this method is immediately obvious: For the shared secret to be established, transmitter and receiver must already have a secure channel available![3]

"Public key algorithms" use a key pair consisting of a public key and a private key. An example is the Rivest-Shamir-Adelman (RSA) algorithm. A key pair is generated once by a computational process, which gives the keys a mathematical relationship such that messages encrypted with the public key can (only) be decoded with the private key and vice versa. The security of public key encryption relies on the fact that it is computationally hard to derive the private key from the public key.[4] A prospective receiver of secure communication would thus go through the following steps:

- Generate a key pair of the desired size
- Distribute the public key to anyone who wishes to communicate with the key owner
- Keep the private key protected, and use it to decode received messages that have been encrypted with the public key

The beauty of public key algorithms is that the private key can also be used to encrypt messages, providing a publicly verifiable signature mechanism: A message encrypted with a private key can only be decrypted with the corresponding public key. This feature is key for the assurance of nonrepudiation: A message that can successfully be decrypted by means of a known public key is known to have been encrypted by the owner of the corresponding private key. The act of encrypting a (hash of a) message with one's own private key is considered a digital signature. A hash (or digest) of a message is a mathematically derived fixed-length string that has a high probability of changing drastically if the input message is changed. The advantage of digitally signing a hash of the message rather than the message itself is a significant time saving and cause less expansion of the data that must be transmitted.

Although public key algorithms are conceptually attractive, they are vulnerable to "man-in-the-middle attacks": A malicious user could dis-

---

[3] There are, of course, numerous ways to securely exchange a secret prior to establishing a data link layer connection: you could meet your correspondent in person and hand over the key on a CD-ROM, you could spell out the key in a phone call, you could involve the postal services... All of these methods are rather inconvenient, especially when the key has to be replaced regularly.

[4] The resources required to do this increase more than proportionally with key size.

tribute a public key under someone else's name, prompting other users to transmit messages destined to the victim, encrypted in a way that allows the malicious user to decrypt them! It is therefore of huge importance that the public keys used in public key encryption can be trusted. For this reason, *certificates* exist. A certificate consists of a public key, the identity of the key owner, and a digital signature from a known, trusted party (there are a number of globally trusted certification authorities whose public keys are supplied with common security-aware software applications such as web browsers).

Port Authentication

Overview

IEEE Std. 802.1X specifies port-based network access control for certain IEEE 802-compliant (bridged) networks. It has been claimed that in order to understand the 802.1X architecture, the best place to start is IEEE 802.11i, which is the derived authentication mechanism for wireless networks. As EFM is not concerned with wireless networks (although PONs have some similarities to wireless), I will not take the scenic route in this section.

For a port to be eligible for the port authentication mechanism, the LAN connected to the port must be a point-to-point LAN. In the case of Ethernet, this excludes the shared-medium port types (10BASE2, 10BASE5, all networks containing a hub) and EPON in LAN-emulation mode. In the case of Wireless LAN (the IEEE 802.11 standards family), a dedicated point-to-point emulation mechanism exists that makes 802.1X port authentication possible.

The basic principle of port authentication is the replacement of the MAC Service access point of the port by two MAC Service access points; one controlled and one uncontrolled. The uncontrolled port is capable of receiving frames as soon as the bridge is switched on (assuming that it is in a "listening" state). This port will be used for the authentication message exchange—initiated either by the attached end station (the *supplicant*), or by the bridge port (the *authenticator*) in response to a non-EAP frame from the end station—which can lead to the activation of the controlled port. The controlled port will be the one that offers subsequent received frames to the MAC Relay via its ISS.

The authentication itself consists of an identification, a challenge, and a response; these are part of the EAP protocol (see below). The user information database used for authentication usually resides outside the switch, in a dedicated *authentication server*. Such servers typically communicate with the bridge via the RADIUS protocol (RFC 2138, RFC 3579), or the Diameter protocol (RFC 3588). The advantage of having a "pass-through authenticator" and a separate authentication server is that the network infrastructure does not have to be upgraded in order to deploy a new authentication protocol.

Extensible Authentication Protocol

The Extensible Authentication Protocol (EAP) was defined in RFC 2284 as an extension to the Point-to-Point Protocol. It supports the same MD5 algorithm as PPP's Challenge Handshake Authentication Protocol (CHAP) for authentication (mandatory), as well as one-time password and token-card authentication (optional). Other authentication algorithms may be added, which is what makes this protocol "extensible." The same mechanisms and architecture are used in IEEE Std. 802.1X to provide authentication in LANs; the associated protocol is named "EAP over LAN" (EAPOL).

The latest RFC on EAP (RFC 3748, currently a "proposed standard") treats EAP in a generic way, regardless of its running over PPP or over a LAN. The fundamental assumption remains that authentication must be possible in situations where network-layer connectivity is not (yet) available.

EAPOL is used in 802.1X to exchange messages between the Port Access Entities (PAEs) in the supplicant and the authenticator, which may lead to the activation of the controlled port. The EAP packet consists of Code, Identifier, Length, and Data fields; on Ethernet networks, it is encapsulated in an untagged or priority-tagged MAC frame, as shown in Table 9.1.

TABLE 9.1 EAPOL Frame Structure (on 802.3 networks)

Field	Length	Description
Destination Address	6 bytes	PAE group address: 01-80-C2-00-00-03
Source Address	6 bytes	Unique MAC address of the transmitting station
Length/Type	2 bytes	PAE Ethernet type = 0x888E (untagged frame) or QTagType = 0x8100 (priority-tagged frame)
QTag Control Information (optional)	2 bytes	If present, the Length/Type field must be set to 0x8100 (QTagType), and a 2-byte MAC Client Length/Type field must follow. Only priority-tagged frames are allowed, so VID must be all-zero.

(continued on next page)

TABLE 9.1   EAPOL Frame Structure (on 802.3 networks) *(Continued)*

Field	Length	Description
MAC Client Length/Type	2 bytes	PAE Ethernet type = 0x888E. This field takes on the role of the Length/Type Field if QTag Control Information is present.
Protocol Version	1 byte	Currently 0x01
Packet Type	1 byte	EAP-Packet=0x00, EAPOL-Start=0x01, EAPOL-Logoff=0x02, EAPOL-Key=0x03, EAPOL-Encapsulated-ASF-Alert=0x04
Packet Body Length	2 bytes	Length of the Packet Body
Code	1 byte	1=Request, 2=Response, 3=Success, 4=Failure
Identifier	1 byte	Field to allow matching of responses to requests (meaning is local to one port)
Length	2 bytes	Length of the EAP packet, including the Code, Identifier, Length, and Data fields
Data	Variable	Format determined by the Code field
FCS	4 bytes	32-bit CRC, as specified for IEEE 802.3 MAC frames

EAP is a "lock-step" protocol, which means that it never has more than one message in flight. This is an important aspect of EAP's robustness against attacks. A (typical) supplicant-initiated authentication exchange consists of the following steps:

1. **EAPOL-START:** The supplicant indicates that it wants to initiate an authentication exchange.
2. **EAP-Request/Identity:** The authenticator requests the identity of the supplicant.
3. **EAP-Response/Identity(myID):** The supplicant submits its identification.
4. **EAP-Request/OTP, OTP Challenge:** The authenticator offers an authentication challenge; this may be a simple request for a password, or a token that must be transformed by means of a known algorithm using the private password as a parameter.
5. **EAP-Response/OTP, OTPpw:** The supplicant submits its password or the response to the challenge.
6. **EAP-Success or EAP-Failure:** The authenticator grants (EAP-Success) or denies (EAP-Failure) access to the network.
7. **EAPOL-LOGOFF:** At the end of the session, the supplicant may request the log off from the network.

A (typical) authenticator-initiated authentication exchange is similar to the supplicant-initiated exchange shown here, except that it lacks message 1. The authenticator may initiate the EAP exchange upon receipt of a non-EAP datagram from the supplicant (e.g., a DHCP autoconfiguration request). The initial non-EAP datagram is dropped; it is recommended that the EAP authentication be completed before the higher-layer protocol's retry timer times out.

Between messages 3 and 4 and between messages 5 and 6, the authenticator communicates with the authentication server to validate the supplicant's requests. By default, the authentication expires after one hour (this duration may be changed by management).

MAC Security

Overview and Definitions

The purpose of MAC Security is to provide encryption facilities (for secrecy and integrity) to a single LAN. If data secrecy and integrity are required between end stations in a bridged LAN, multiple secure hops may be combined to provide an end-to-end association; in a provider-bridged network, the customers' end-to-end secure transmission is relayed transparently through the provider network. MAC Security does not protect against traffic analysis (addresses are not encrypted), nor does it provide nonrepudiation (there is no protection against source address spoofing, other than the assurance that the source address was not changed between the transmitter and the receiver).

To reach the goals of secrecy and integrity, the MAC Security standard introduces a "Security Entity" (SecY), which resides between the MAC's original ISS and the MAC Relay (in other words, it offers a new ISS access point to the MAC Relay, on behalf of the underlying MAC). The Security Entity and its peer select a Cipher Suite through the Key Agreement Entity (KaY), and use this Cipher Suite for subsequent encryption activities.

A Cipher Suite is a selection of algorithms for authentication and encryption that are known to work well together. This avoids the introduction of a security weakness in the network through the selection of algorithms for authentication and encryption that would reduce each other's reliability. The current draft of the standard requires that at least one Cipher Suite be supported that authenticates but does not encrypt the user's data (GMAC), and a Cipher Suite that authenticates *and* encrypts the user's data (GCM-AES). The Security Entity must additionally support the *Null Cipher Suite*, which does neither.

The secure equivalent of a LAN is a Connectivity Association (CA). This is a set of ports connected to a single physical LAN that are able to communicate with each other by means of a Secure MAC Service instance. The same physical port may be part of different Connectivity Associations (different subsets of the same LAN), identified by different Secure MAC Service access points.

The Secure MAC Service access points rely on a Secure Channel (SC) to transmit frames to all other members of the CA. The Secure Channel, in its turn, relies on a Security Association (SA), which uses the agreed Cipher Suite with a particular set of keys and which is limited in time; a sequence of Security Associations (overlapping in time) provides for continuity of the supported Secure Channel.

Readers familiar with IPSec, a suite of protocols that offers security at the IP layer (RFC2401), may notice that the service offered by MAC Security are similar to the ones offered by the Authentication Header (AH) and Encapsulating Security Payload (ESP) protocols. This is true, but the fact that the latter are layer-3 protocols makes them suited for a different set of applications than MAC Security. Where IPSec is deployed between IP hosts (or between an IP host and an IP gateway), MAC Security can be used on LANs that do not have any IP-awareness (e.g., an EPON where all end points are bridge ports). Where security is only required on certain segments of a bridged LAN, MAC Security can offer this service without burdening the user of the end-to-end connection with security management and without the need to install IP gateways in the network.

MACsec Protocol

The MACsec protocol introduces the SecTAG, a new tag between the source address and the encrypted payload.[5] A 128-bit Integrity Check Value (ICV) is appended to the encrypted payload. These operations (SecTag, encryption,[6] and ICV) expand the frame size significantly. For a frame whose original (unencrypted) payload is close to the maximum of 1500 bytes, the resulting encapsulated MACSec frame would exceed

---

[5] In this case, the "payload" includes the original Length/Type field and optional VLAN information (C-TAG/S-TAG).

[6] Some encryption algorithms, such as the default cipher suite GCM-AES, produce the exact same number of bytes of encrypted text as the original cleartext; others introduce cryptographic expansion.

the maximum size of 1522 bytes (not including preamble and SFD). To allow such frames to be transmitted on future LANs, the IEEE 802.1 Working Group has requested a revision of the maximum frame size of 802.3, much in the same way that the frame size was increased by 4 bytes for the introduction of VLAN tags. The request is being considered by a dedicated Task Force of the 802.3 Working Group; most likely, there will be a more future-proof expansion of the MAC frame, which can also accommodate future requests (see also Chapter 10).

The SecTAG is composed of following fields:

Tag Control Information (TCI).   This field conveys the version number of the MACsec protocol, it indicates whether the SCI is implied by the MAC Source Address or explicitly expressed, it indicates the use of the EPON single-copy broadcast capability, and it indicates the use of the Short Length field.

Association Number (AN).   This field identifies up to four different Secure Associations within a Secure Channel.

Short Length (SL).   This field is set to the number of octets in the Secure Data field, if that number is less than 64, to avoid misinterpretation of padding bytes as data. Otherwise, SL is set to zero.

Packet Number (PN).   This feature protects against "replay attacks," in which the attacker captures a conversation between two parties, and replays one side of the conversation to the other party to maliciously act on behalf of the first party. If the receiver checks the PN of frames coming in on a particular CA, it can easily detect out-of-sequence packet exchanges, which indicate that packets are being replayed.

Secure Channel Identifier (SCI)—optional.   This identifier is used on LANs that support multiple CAs to exist at the same time (shared LANs). It allows the receiver to identify the appropriate Cipher Suite and key set for the incoming frame and present the MPDU to the appropriate Secure MAC Service access point.

The PN is used as a *nonce*; this is an initialization value used by the encryption algorithm in addition to the key and the cleartext in order to generate unique ciphertext. The destination address and source address of the MAC frame are used as additionally authenticated data (AAD), which means that they will be protected by the ICV, but not be encrypted.

TABLE 9.2   MACsecPDU Structure

Field	Length	Description
Destination Address	6 bytes	Unique MAC address of the receiving station
Source Address	6 bytes	Unique MAC address of the transmitting station
Length/Type	2 bytes	MACsec Ethertype (to be assigned by IEEE)
SecTAG	1 or 2 bytes	Security TAG
Secure Data Type	n bytes	Encrypted Payload
ICV	8 bytes	Integrity Check Value
FCS	4 bytes	32-bit CRC, as specified for IEEE 802.3 MAC frames

Use of MAC Security in EPON

MAC Security was designed to offer security and privacy in shared infrastructures such as EPON. Provisioning the Secure MAC Service as a bidirectional point-to-point MAC Service between the OLT and each of the ONUs effectively gives the EPON the same level of security as a set of physically separate point-to-point links. At the OLT, either the SCI or the MAC address of incoming frames can be used to determine from which ONU the frame originated and which Cipher Suite and key set must be used to authenticate and decrypt the frame.

The use of the Single-Copy Broadcast mechanism at the OLT (see the section "Broadcast and LAN Emulation in PONs" in Chapter 7) can also be secured by MAC Security. Although the MAC Service definition does not support instances that offer only unidirectional point-to-multipoint connectivity, the definition of a Secure Channel does map nicely to the concept of a Single-Copy Broadcast MAC. It is possible to define a set of Connectivity Associations (one for each ONU) in such a way that each ONU can securely receive the downstream broadcast data. The Secure MAC Service that is provided in this case remains unusual in the sense that it does not support bidirectional data traffic.

Key Exchange

A separate protocol, KeySec, is being developed to generate keys for the different Secure Associations. This protocol does not perform authentication, but assumes that the different stations in the Connectivity Association have been preauthenticated in some way. The current draft of the 802.1af standard (an amendment to IEEE Std.

802.1X) assumes that the stations have a preshared secret key (exchanged for example through EAP), a public/private key algorithm, or distribution outside the network.

Although the MACsec protocol supports the use of a different set of keys for each Secure Channel, the KeySec protocol generates a common Secure Association Key (SAK) for all the members of a Connectivity Association. As a result, each station can participate in a shared secure LAN storing only as many keys as would be required for a secure point-to-point connection. As the transition from one key to the next does not happen simultaneously in all secure channels, the stations have to keep an old key and a new key active until the last secure channel has migrated.

The SAK is calculated as a pseudo-random function from the shared key and a set of key contributions from all participants. The inclusion of the preshared secret key ensures that an attacker who captures the key contributions from the different stations is still unable to calculate the SAK used by the preauthenticated stations.

10

Beyond the
First Mile

*"Service providers are on the lookout for
supporting technologies that enable newer
Ethernet-based services such as transparent
LAN service (TLS) connecting various
customer sites across a metropolitan domain
(metro Ethernet)."*
—GIRISH CHIRUVOLU ET AL. [13]

Aggregating Subscriber Traffic

If the benefits of Ethernet in terms of cost and ease of deployment are
obvious for the first mile, because the Ethernet in the First Mile (EFM)
technology was *designed* to be cost-effective and simple, this is not necessarily the case for the aggregation network, the layer-2 network that
provides connectivity between the subscribers of a certain service
provider and a Broadband Access Server (BRAS) or an IP service edge
that interfaces with the IP core network. The technology for the aggregation network can be selected independently of the technology in the
first mile, provided that there is adequate protocol conversion at the
end of the first mile (remote unit, DSLAM, access multiplexer).

The functions of the aggregation network are unavoidably more
complex than those of the first mile. The first mile is a simple point-
to-point connection (or virtual point-to-point connection, in the case of
EPON), taking in data on one end and handing it to the client at the
other end. The aggregation network is a complex topology where
measures must be taken to use the available bandwidth as efficiently

as possible while avoiding congestion, support different levels of Quality of Service (QoS), and prohibit potentially harmful traffic from malicious subscribers to reach other subscribers' equipment; in brief, it must deliver a satisfactory quality of experience for the end user.

Ethernet technology *can* be used in the aggregation network, as Ethernet is obviously no longer merely a *local* area network (LAN) technology. The reach of the more recent Ethernet Physical Layer entity sublayer (PHY) is in the order of tens of kilometers, and with Ethernet switches, separate segments can be combined to create Ethernet networks of huge proportions. However, the LAN communication paradigm—everyone can see all traffic—is neither appropriate nor sufficiently scalable for networks that are used for the commercial deployment of a network service. The low cost of Ethernet LAN equipment is therefore no longer a given when the technology is extended to operate in an aggregation network. The fact that Ethernet's physical layer solutions are well understood and relatively cheap does remain attractive.

A number of technical drivers favor an evolution toward Ethernet in the aggregation network. A large portion of the traffic that is being handled in aggregation networks includes Ethernet somewhere in its protocol stack, as about 90 percent of IP traffic originates from Ethernet LANs [12]. This implies that removing the lower layers (the encapsulation around the Ethernet frame) will make the transport more efficient. Furthermore, Ethernet handles multicast traffic very well, because the duplication of frames towards multiple ports of a network node is a basic function of the bridging paradigm. The IP protocol suite, by design, works perfectly on top of Ethernet; as IP-based services become more prevalent, this is another reason to build the aggregation network with Ethernet blocks.

This chapter discusses the main issues related with the subscriber access network beyond the first mile and the network nodes and protocols that are available to deal with these issues. The basis of all modern full-duplex Ethernet networks with more than two stations is a bridged LAN. Current standards and draft standards provide three levels of virtualization on top of MAC-bridged LANs: virtual bridged LANs, provider-bridged networks, and provider backbone networks.

Bridged LANs

The two different modes of operation of the Ethernet MAC and PHYs lead to two distinct categories of Ethernet LANs: half-duplex LANs, which may have multiple stations attached to them (each connecting directly to a shared medium, such as coaxial cable; or indirectly, via a repeater or hub), and full-duplex LANs, which have exactly two stations on a single full-duplex point-to-point link (e.g., two PCs equipped connected NIC-to-NIC with a cross-over cable).[1] Today's practical networks belong to neither of these categories; they are "bridged LANs" consisting of multiple (typically full-duplex) LANs interconnected by means of a bridge.[2]

Principles of Transparent Bridging

Bridging is said to happen "transparently" if the attached stations have no way of distinguishing between being connected to a single LAN and being connected to a bridged LAN. Transparency is achieved by assuring rapid forwarding of all incoming frames toward the other LANs attached to the bridge, with minimal probability of frame loss, and with no frame duplication or reordering[3] occurring in the process.

A bridge is different from a repeater or hub in that it does not necessarily repeat all incoming frames on all other ports. It remembers the source addresses of incoming frames and the ports on which they came in, and forwards further frames to the appropriate port only, a property known as "dynamic filtering." Only if a frame comes in with a destination address that has not yet been learned or statically configured by management, or if the learned information has been removed due to expiry of the ageing timer (typically after 5 minutes), the frame is "flooded" to all ports in a repeater-like fashion.

The filtering properties of a bridge and the fact that a bridge can buffer incoming frames internally before forwarding them—implying that frames don't have to be forwarded in near-real time—give the bridge another important advantage over the repeater: A bridge can interconnect full-duplex LAN segments, while a repeater only serves to bring together half-duplex LANs.

Bridged LANs, like shared LANs, operate under the assumption of MAC address uniqueness. If multiple end stations in a bridged LAN have the same MAC address (due to an error in configuration, faulty equipment, or malicious intent), the forwarding process cannot offer

---

[1] This is a cable that connects one end's transmit interface to the other end's receive interface.

[2] The term "switch" or "Ethernet switch" is often used to designate a (hardware) implementation of a bridge; a "switch" may, however, contain other functions in addition to the pure bridging function.

[3] The reordering criterion is only relevant with respect to frames sharing the same source and destination addresses, and the same priority class.

correct connectivity to all such stations. This is one of the reasons why the pure "bridged LAN" model is not suitable for the transport of subscriber traffic.

As described, bridges achieve transparency in an efficient manner; the MAC Service is preserved, such that an end station need not know whether its peer is on the same LAN or behind a bridge, and yet data are forwarded only to those segments that really need to receive it. This situation is radically different from Token Ring LANs, in which end stations need to be aware of the topology of the bridged LAN they are connected to. Token Ring frames are source-routed; they leave the end station with the necessary information to find their way through the network. Although Ethernet frames technically also have the possibility to carry source routing information, this option is never used in pure Ethernet networks (it may, however, be used in hybrid Ethernet-Token Ring networks [37]).

A bridged LAN may contain one or more bridges, deployed in a star (or tree) topology. This topology corresponds very nicely to that of a subscriber access network. In a typical Ethernet subscriber access network, all customer equipment connects to an EFM modem via traditional point-to-point Ethernet links, all modems are in turn connected to an Access Multiplexer via EFM links, and all Access Multiplexers are further connected to a layer-3 gateway via a cascade of Ethernet switches. The EFM modem, the Access Multiplexer, and the Ethernet switches are all boxes that contain bridging functionality. As will be seen in the next section, pure bridging functionality is not sufficient to build a secure, reliable, and revenue-generating access network.

Spanning Trees

Complex interconnections of bridges and LANs may physically create closed loops that could severely disturb the operation of the bridged LAN if no special precautions are taken: Frames could end up being forwarded forever, or arriving at the same bridge via two different ports, causing the learning process to go awry. Bridges run a spanning tree protocol[4] to exchange topology information in order to detect such loops and break them by blocking certain ports.

The original Spanning Tree Protocol (STP), as defined in the 1998 edition and older editions of IEEE Std. 802.1D, operates by electing a

---

[4] A tree is a loop-free graph; a spanning tree is a fully connected tree, i.e., a tree that contains all the nodes of the original graph.

unique root bridge in the network, then distributing among all bridges the cost associated with each path towards the root bridge and allowing only the "lowest-cost path" from any bridge to the root to remain active. On any given bridge, the only ports that are in the forwarding state are the "root port" (the port that is connected to the lowest-cost path to the root) and "designated ports" (ports that are connected to a LAN for which the lowest-cost path to the root runs through that bridge).

The root path cost is calculated as the sum of the costs of all the network segments making up the path. The cost attributed to each segment can be configured through the bridges' management, but it is recommended by the standard to use a decreasing function of the bitrate supported by the segment, such that fast links have a lower cost than slow links, and thus get priority for activation.

The automatic activation of a spanning tree by the STP allows a network operator to provision redundant connections as back-up links, which will only become active when the original link becomes inactive (this could happen when the original link goes down due to a PHY or media failure, or when it is taken out for maintenance). The protocol also makes sure that dynamic filtering entries that have become unreliable due to a topology change are removed from the bridge's learning tables.

STP requires the network diameter (the maximum number of bridges between any two end stations) to be less than or equal to 7. Since the original STP was developed, two new spanning tree protocols have been introduced to fix this and other shortcomings of STP.

Rapid Spanning Tree Protocol (RSTP). RSTP, specified as IEEE Std. 802.1t and now part of IEEE Std. 802.1D-2004, has better convergence properties after a failure or reconfiguration in the network. RSTP configures inactive bridge ports as "back-up ports" (ports that can take over the role of a failing designated port) or "alternate ports" (ports that can take over the role of a failing root port), allowing those ports to rapidly transition from the discarding state to the forwarding state in the event of a topology change. For this purpose, it uses knowledge of ports attached to only one other bridge or to an end station. Forwarding tables are moved to the activating port when appropriate. The increase in rapidness comes at the price of very a small probability of frame duplication or reordering.

Multiple Spanning Tree Protocol (MSTP). MSTP supports the co-existence of different spanning trees on the same network. The MSTP protocol divides the network in different regions, which are interconnected by a

Common Internal Spanning Tree (CIST). Within each region, different Multiple Spanning Tree Instances (MSTIs) can be active. Different VLANs (see below) can be put on different MSTIs. MSTP is designed to work when there are portions of the network that use STP or RSTP.

All the STPs mentioned here require a minimum amount of communication between the bridges in the network. Bridge Protocol Data Units (BPDUs) are used for this purpose, which are transmitted as LLC-encapsulated Ethernet frames to a special reserved address.

MAC Bridges

In general, bridges are not limited to Ethernet LANs. "Media Access Control (MAC) Bridges" as specified by IEEE Std. 802.1D, are capable of translating between different MAC technologies, including different frame types and CRC algorithms. The generic diagram of a MAC bridge is shown in Figure 10.1.

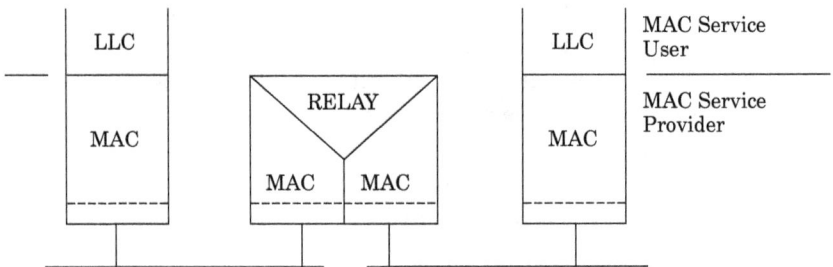

Figure 10.1   Generic MAC bridge diagram.[5]

A MAC Bridge consists of a MAC Relay function, which interconnects a number of MAC Entities. MAC Entities are instances of the MAC sublayer for a particular technology—henceforth always assumed to be Ethernet—which present an ISS interface[6] to the MAC Relay, and a MAC Service interface to an optional management entity and the so-called "bridge brain."

The ISS interface is drawn as an oblique line in Figure 10.1 and Figure 10.2, while the MAC Service interface is drawn as a horizontal line in Figure 10.1 and Figure 10.2. Both figures are taken from the respective bridging standards and are typical graphical representations

---

[5] From IEEE Std. 802.1D-2004, "Media Access Control (MAC) Bridges," by IEEE. © 2004 IEEE. All rights reserved.
[6] The ISS and MAC Service are explained in Chapter 1.

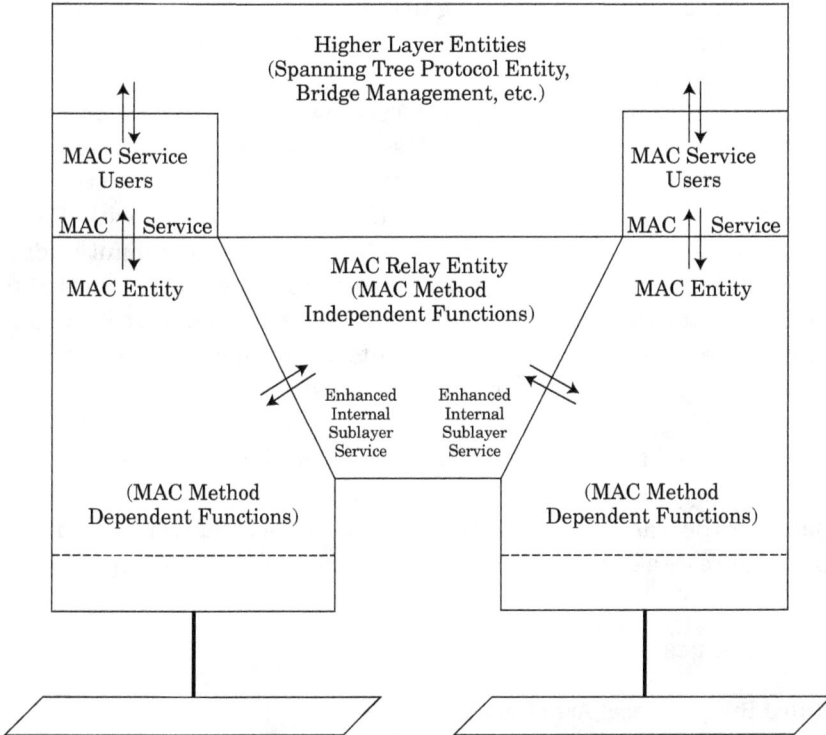

Figure 10.2 VLAN-aware bridge diagram.[7]

of the MAC Relay. The advantage of this representation is that it clearly shows the position of the different interfaces in the system. The disadvantage of this representation is that it does not allow visualization of more than two ports on a bridge, without resorting to a sort of three-dimensional rendering. The reader shall bear in mind that the general model of these figures applies to bridges with any number of ports.

Frames received on a particular port will be presented to the MAC Relay via the ISS interface, except those frames that have as destination address the address that is associated with the port's management entity or a bridge brain function. Frames destined for the management entity will be presented to that entity via the MAC Service interface. Based on its forwarding rules, either learned or configured by management, the MAC Relay will forward the frames it receives to other ports via these ports' ISS interface. MAC frames will

---

be forwarded or filtered according to their Destination Address alone—other information such as the Length/Type field or the user_priority are *not* taken into account.

Existing and future layer-2 protocols, which by their nature must never be forwarded beyond the link segment on which they originated (spanning tree protocols, slow protocols, etc.), use special destination addresses from a range of reserved addresses (all starting with 01-80-C2-00-00-0?), and are never forwarded by a standard compliant bridge.

Broadcast frames and multicast frames are by default forwarded to all ports. Individual multicast addresses can, however, be restricted to a set of registered ports. This is particularly useful when the multicast mechanism is used to deliver high bitrate streaming multimedia content to (paying) subscribers.

Unlike IP addresses, MAC addresses do not have an inherent hierarchy to support "subnetting." Forwarding decisions in a bridge are made on the basis of a complete match (or lack thereof) of a 48-bit MAC address, as opposed to the longest prefix match used in IP.

VLAN Bridges

Virtual Bridged Local Area Networks

Different *logical* topologies may be deployed simultaneously and separately on a single *physical* bridged LAN by means of Virtual Bridged LAN (VLAN) tagging, which is governed by IEEE Std. 802.1Q. This means that frames belonging to one VLAN will be forwarded according to the forwarding rules of that VLAN, such that they can be "contained" within a particular region of the network, preventing them from reaching end stations that do not belong to the same VLAN. This containment also provides separation of broadcast domains, such that broadcast and multicast frames do not use more resources than necessary.

The 16-bit VLAN tag (shown as QTag Control Information in Table 1.1) consists of a 12-bit VLAN ID, a 3-bit priority, and a 1-bit CFI field. The tag space allows 4094 different VLAN IDs; VLAN ID value 0xFFF is reserved, and a frame with a zero VLAN-ID is not considered VLAN-tagged, but only priority-tagged. The priority field may be used for queueing purposes in a bridge, to ensure that higher-priority frames are transmitted to the destination ports first in the case of congestion; the Ethernet MAC itself does not observe QoS indications.

The VLAN tag or QTag was not originally part of the Ethernet frame; this field was added to the 802.3 frame on request of the IEEE

802.1 Working Group during the preparation of the VLAN standard. This caused an increase of the maximum MAC frame size from 1518 bytes to 1522 bytes (not including preamble and SFD).

A VLAN-aware bridge assigns a VLAN ID to incoming frames and forwards them according to the forwarding table associated with that VLAN, which can contain learned and statically configured entries. The VLAN ID is assigned in one of the following ways:

- If the frame carries a VLAN tag, the VLAN ID in the tag is assigned to the frame.
- If the frame does not carry a VLAN tag, the default port VLAN ID (PVID) is assigned to the frame.
- If the bridge supports the optional port-and-protocol based tagging capability, a VLAN ID may be assigned to the frame on the basis of the incoming port *and* the MAC Client protocol indicated in the type field.
- If the bridge supports other (proprietary) tagging criteria, these may be used to assign a VLAN ID.

It is important to note that the bridge treats the different virtual LANs as truly separate entities—it behaves, in fact, as though it were multiple bridges. As explained, VLAN IDs are assigned to individual frames entering a bridge and are used for the selection of the appropriate forwarding table. The fact that the frames are assigned a VLAN ID does not necessarily mean that they will be *tagged* on the outgoing port—this is a managed property of the individual ports. These different rules for VLAN bridging allow the network operation to configure the links attached to a VLAN bridge in a variety of ways.

Access links.   Access links are links that carry untagged or priority-tagged traffic belonging to a single VLAN. A VLAN ID is assigned to the frames as they enter the VLAN bridge to which the link is attached, depending on the port of entry (PVID). The traffic from different users, subscribers, departments, or other entities may be aggregated onto a VLAN-bridged backbone network by means of access links; this removes the need for any VLAN-awareness at the end stations.

Trunk links.   Trunk links are links that carry VLAN tagged traffic belonging to different VLANs. These links are the portions of the VLAN bridged network that are actually *shared* between VLANs (as opposed to access links, which only serve a single VLAN). As all the traffic carries VLAN tags, it is easy to keep traffic from different users logically separate.

Hybrid links.   Hybrid links are links that do not belong in the two categories above. They may carry a combination of untagged, priority-tagged, and VLAN tagged traffic, according to the specific needs of the network.

VLANs can provide a certain level of separation between different parts of a network, by confining traffic to only those ports that belong to the same VLAN. In this way, a number of attached end stations can be prohibited from communicating directly with each other, without first passing through a different node (e.g., to be passed through a layer-3 gateway, if the layer-3 protocols support this type of connectivity).

Nevertheless, VLANs alone must not be relied on for security. The VLAN has no means of stopping a malicious user who gains physical access to a network segment that is a member of the VLAN (or a legitimate user with bad intentions) from intercepting and reading the frames on that segment. A malfunctioning or poorly configured bridge may forward traffic to ports on which it does not belong, again without any additional protection from interception. If secrecy is an important concern in the network, additional measures must be taken (see Chapter 9).

VLANs, like bridged LANs, operate under the assumption of MAC address uniqueness. However, in a well-designed network, address space collisions between VLANs will not affect the normal operation of the individual VLANs, as the forwarding tables of the different VLANs can be kept strictly separate.

Using VLANs

VLANs are used in private networks to segregate traffic belonging to different administrative entities, different IP subnets, or different applications. If a judicious combination of access links and trunk links is deployed, the VLAN-aware infrastructure can remain transparent for the end users: It's there, but the end user equipment operates in the same way as if it were not there. Only the core of the network needs to deal with tagged traffic.

The corporate situation could be compared to the situation of a public network operator. Operators use their networks to offer services to customers who expect certain service level guarantees. Shared LANs, and to a lesser extent bridged networks, have to divide their resources between the different connected stations, and are therefore poorly suited to offer these guarantees. VLANs can provide a degree of segregation of traffic belonging to different customers, and improve the quality of the service offered to the customers.

In traditional ATM-based subscriber access networks, services are provided to individual subscribers by means of data pipes (virtual circuits) with well-defined average and maximum bitrate characteristics. Although traditional Ethernet lacks the capability of setting up virtual circuits, it is logical to think of VLANs as a way of introducing some of the carrier-gradeness of ATM to Ethernet. Although VLANs do not define circuits, they do allow an operator to set up separate subspaces with independent forwarding rules and independent spanning trees in a single physical network, thus offering means to perform elementary traffic engineering.

Digital Subscriber Line Access Multiplexers (DSLAMs) based on a VLAN-aware bridge, operating as a DSL-line–VLAN cross-connect, have been used in small-scale DSL deployments. The most obvious problem associated with using VLANs in a subscriber access network is scalability. An Ethernet subscriber access network using VLANs to strictly segregate subscriber traffic (e.g., by assigning a different port-based VLAN ID to all incoming subscriber traffic) would support only 4094 subscribers, which is a number that doesn't make economic sense. Furthermore, it is not clear how to support traffic that is already VLAN-tagged by the customer—typically a corporation wishing to transport its own tagged traffic transparently over the provider network—in such a subscriber access network. This shows that using VLANs as a substitute for virtual channels is going to be a stretch. These and other problems are addressed by IEEE P802.1ad "Provider Bridges," a new amendment to IEEE Std. 802.1Q, which is currently being developed. The central theme in this amendment is "VLAN stacking."

Provider Bridges

Provider Bridged Networks

The central component of a Provider Bridge is a *Service VLAN-aware Bridge component*, which essentially does the same thing as a VLAN bridge, as shown in Figure 10.2, but it operates on a different tag. The Service VLAN-aware Bridge component can add or remove Service VLAN (S-VLAN) tags independently of the original VLAN tag (now called Customer VLAN or C-VLAN, to distinguish it from the S-VLAN), to create a second level of traffic segregation outside the control of the subscriber.

C-VLAN tagged data from the customer network can be "tunnelled" transparently through the provider network. Certain layer-2 bridge protocol datagrams that would be filtered by a customer bridge or non-VLAN bridge (e.g., for spanning tree configuration) can also be forwarded transparently by a provider bridge, because provider bridges use a different set of reserved addresses for which forwarding is prohibited.

In addition to the Service VLAN-aware Bridge component, a Provider Bridge may include one or more Customer VLAN-aware Bridge components attached to Service VLAN-aware ports. These Customer VLAN-aware Bridge components operate *exactly* like a VLAN bridge, operating on the original (C-VLAN) tag. As a result of this two-in-one bridge architecture, the ports of a Service VLAN-aware Bridge component can be divided in three different categories, shown schematically in Figure 10.3.

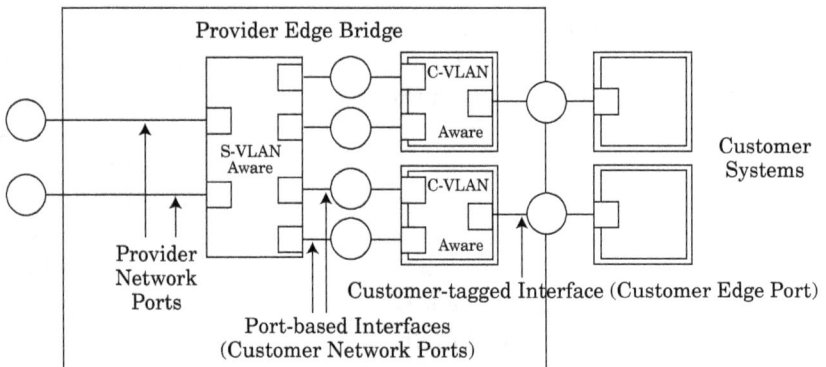

Figure 10.3 Naming of the ports on a provider bridge.[8]

Provider network port.   This is a port attached to the Service VLAN-aware Bridge component, and connecting to other provider equipment.

Customer network port—connecting to customer.   This is a port attached to the Service VLAN-aware Bridge component and connecting to customer equipment.

---

[8] From IEEE Draft P802.1ad/D3.0, "Provider Bridges," by IEEE. Copyright 2004 IEEE. All rights reserved.

Customer network port—connecting to C-VLAN–aware Bridge component.
This is a port attached to the Service VLAN-aware Bridge component
and connecting to the Customer VLAN-aware Bridge component (multiple ports may interconnect the same two Bridge components). The
Customer VLAN-aware Bridge ports actually facing the customer are
called Customer Edge Ports.

The mechanism to indicate the presence of an S-VLAN tag, and its
location in the Ethernet frame are still to be determined; the former by
the IEEE 802.1 Interworking Working Group, the latter by the IEEE
802.3 CSMA/CD Working Group (undoubtedly leading to an increase
in the maximum Ethernet frame size by another 4 bytes). The architecture specified in the current draft of the standard suggests that an
S-VLAN tagged frame will have a new Ethertype value in its
length/type field, followed by the 2-byte S-VLAN tag, followed by the
"payload," which may include a QTag identifying a certain C-VLAN.

Existing proprietary implementations of provider bridges use a second QTag to carry the S-VLAN information, hence the names "VLAN
stacking" and "Q-in-Q." Earlier on in the project, an alternative
scheme to improve scalability was proposed, based on the encapsulation of Ethernet frames inside other Ethernet frames, the latter being
wholly under the control of the provider. This proposal was informally
known as "MAC-in-MAC," and will be under discussion again in the
recently started project on "Backbone Ethernet" (IEEE P802.1ah).

Using S-VLANs

The use of the S-VLAN tag to segregate traffic belonging to different
subscribers resolves the problem of transporting frames that have
already been C-VLAN tagged by the subscriber. Giving the service
provider a different tag to work with does not solve the scalability
issue outlined above. The Service VLAN space is just as small as the
Customer VLAN space. Only when the service provider assumes control over *both* the Customer VLAN tag (when it is not used by the subscriber) and the Service VLAN tag, a 24-bit tag space is created which
can uniquely label 2^{24} (over 16 million) virtual data streams.

The two labels can be assigned by different provider-owned bridges;
the first (C-VLAN) tag is, for example, assigned by a smaller access
node such as a remote unit in an MxU or a cabinet, acting as a
Customer Bridge, and the second (S-VLAN) is assigned by an aggregating switch deeper inside the network, acting as a Provider Bridge.

Alternatively, a single access node can combine the functions of a Customer Bridge and a Provider Bridge and keep the entire combined C-VLAN/S-VLAN space at its disposition. Both scenarios extend the concept of the VLAN cross-connect (see below), which attempts to translate the notion of ATM virtual circuits to the Ethernet world.

Provider bridged LANs still operate under the assumption of MAC address uniqueness. As is the case for VLANs, address space collisions between S-VLANs will not affect the normal operation of the individual S-VLANs in a well-designed network. However, the protective walls around the C-VLANs disappear in the S-VLAN–aware bridge component: A forwarding table will be selected on the basis of the S-VID alone, so the occurrence of duplicate addresses in different C-VLANs that come together in the same S-VLAN *will* cause a disruption of the connectivity for the concerned end stations.

The 2-in-1 Bridge Model used to specify Provider Bridges introduces a number of tagging schemes that were previously impossible to deploy in a single standard-compliant bridge. To understand how this works, it is important to notice that each Customer VLAN-aware Bridge component may be connected to the Service VLAN-aware Bridge component through multiple Customer Network Ports (implemented as *internal* EISS interfaces). The Customer VLAN-aware Bridge component makes forwarding decisions in the same way a VLAN bridge would: For each frame, the forwarding rules that apply to the frame's Destination Address are looked up in the table associated with the frame's C-VLAN identifier (C-VID).

As a result the frame can be forwarded to one or more Customer Edge Ports (leaving the Provider Bridge) and/or to one or more Customer Network Ports (going into the Service VLAN-aware Bridge component). Upon ingress in the Service VLAN-aware Bridge component, the frame may receive an S-TAG based on the ingress port—in this case, determined by the forwarding decision in the Customer VLAN-aware Bridge component. The following possibilities result from this architecture.

C-TAG based S-VLAN tagging.   A frame coming in on a Customer Edge Port, can be forwarded to a specific Customer Network Port, based on the frame's C-TAG. In the S-VLAN-aware Bridge component, a port-based S-VID can be assigned. This two-step setup effectively translates sets of C-VIDs into specific S-VIDs. In this way, the customer can get some control over the handling of his traffic in the service provider's network by transmitting frames with appropriate C-TAGs.

DA-based S-VLAN tagging. A frame coming in on a Customer Edge Port, can be forwarded to a specific Customer Network Port, based on the frame's Destination Address. In the S-VLAN-aware Bridge component, a port-based S-VID can be assigned. This two-step setup effectively translates sets of Destination Addresses into specific S-VIDs. It is useful when using S-VLANs to segregate traffic for different services, which is known to pass through fixed points in the network (e.g., a router or a broadband access server). It does not require any VLAN-awareness at the customer side.

Backbone Networks

In deployments where the customer retains full control over the entire C-VLAN space, as is the intent of the Provider Bridges standard, the provider faces the original scalability problem again. In this case, other technologies must be used to securely "tunnel" traffic belonging to different customers through the network. ATM, Frame Relay, and Multi-Protocol Label Switching (MPLS)/Virtual Private LAN Service (VPLS) are existing possible solutions [20]. The recently started Provider Backbone Bridges project (IEEE P802.1ah) is developing an all-Ethernet solution to the same problem.

Multi-Protocol Label Switching

Multi-Protocol Label Switching (MPLS) was developed as a connection-oriented packet-based layer-2.5 solution to add traffic engineering and QoS capabilities to the IP/Ethernet protocol stack. MPLS is used "between" Ethernet and IP; that is, the IP packet is encapsulated in an MPLS frame, which is in turn encapsulated in an Ethernet frame. The packet undergoes MPLS switching from one router to the next: At every intermediate MPLS switch, forwarding decisions are taken on the basis of the *MPLS label*. Between MPLS switches, the frame undergoes Ethernet switching: At every intermediate Ethernet switch, forwarding decisions are taken on the basis of the S-VLAN and/or C-VLAN tags (if present) and the destination MAC address.

At the edge of an MPLS network, incoming frames are classified into a Forwarding Equivalence Class by a Label Edge Router (LER), based on any relevant layer-3 and layer-4 information. The Forwarding Equivalence Class maps to a unidirectional Label Switched Path (LSP), which carries the frame from one Label Switch Router (LSR) to the next. The LSP is set up before any data are transmitted in one of two ways: It either uses hop-by-hop routing, where each LSR independently selects the next hop for the Forward Error Correction sublayer (FEC) by swapping labels, or explicit routing, where the entire path information is stored in the frame from the start as a stack of labels. Routing protocols such as BGP and RSVP can be used for label distribution.

With MPLS, operators can easily route traffic around bottlenecks, congested areas, and network failures. MPLS further provides end-to-end OAM, Quality-of-Service, resiliency, and failure indications. Like ATM, MPLS "tunnels" data flows through the network, ensuring that different flows cannot affect each other, a property that is used to create Virtual Private Networks (VPN) or Virtual Private LAN Service (VPLS).[9] The latter application changes the traditional protocol stack: A layer-2 protocol is transported over a layer-2.5 protocol, which is itself transported over a layer-2 protocol (Ethernet-over-MPLS-over-Ethernet), which is known as "Martini encapsulation" [12].

Provider Backbone Bridging

The IEEE P802.1ah Provider Backbone Bridging project further explores the MAC-in-MAC solution, which was the rejected candidate for the Provider Bridging standard. The solution relies purely on the Ethernet switching paradigm (forwarding frames on the basis of their destination MAC address), but enhances it by defining two distinct and independent address spaces: the edge address space and the backbone address space.

Upon entering the first provider backbone switch, the Ethernet frame—still a regular Ethernet frame at this stage, which may or may not carry a C-VLAN and/or S-VLAN tag—is encapsulated in another Ethernet frame: This means that a new destination address, source address, and type[10] field are prepended to the frame. The switch at the edge of the provider backbone network must therefore have capabilities similar to that of an LER, to assign outer frame parameters to the frame based on the available layer-2, layer-3, and layer-4 information. Knowledge about potential destinations must be present in the edge switches. However, subsequent bridges operate by the self-learning mechanisms that are also used by regular MAC bridges.

---

[9] Also known as Transparent LAN Service (TLS).
[10] A *type* field indicating that the payload of this Ethernet frame is an Ethernet frame; such an Ethertype did not exist to date.

The MAC-in-MAC solution solves two of the problems that we have run into before.

- The *scalability* of this solution far exceeds that of the VLAN-based solutions, as the 48-bit MAC address fields can identify 2^{48} distinct enitities, while the actual number of edge nodes will be relatively small. Within each provider backbone bridge conversation, millions of end-to-end conversations can be transported while the provider backbone bridge only needs to be aware of the MAC addresses of its peers.
- The *uniqueness* of MAC addresses used by subscribers is no longer an issue as far as frame forwarding in the provider backbone network is concerned; provider backbone bridges only look at the outer destination address, which always designates provider-owned equipment, the MAC addresses of which can be managed by the provider. Note that collisions in the subscriber MAC address space may still impact the identification of the Forwarding Equivalence Class at the edge switch.

A MAC-in-MAC frame is a bit like Martini encapsulation without the MPLS in between. Whenever Ethernet frames carry a protocol stack that contains another instance of Ethernet, the matter of the maximum frame size comes up: By definition, a complete maximum-sized Ethernet frame cannot fit into the payload field of an Ethernet frame. This is a *logical* problem that cannot be avoided with protocols that have a maximum frame size.[11] A generic solution is therefore impossible.

What is possible is a specific modification to the frame format, which allows a second {SA, DA, Ethertype} set in front of the original frame header, and an increased maximum frame size. To be more future-proof, the IEEE 802.3 Working Group is currently investigating changing the Ethernet frame format by adding variable-size "envelope" fields at the beginning and the end of the frame, accommodating the extra fields required for MAC security, provider bridges, provider backbone bridges, and any future protocol extensions.

---

[11] As networking hardware has finite memory, protocols must have a maximum frame size. If the higher-layer protocols want to transmit larger datagrams, they must provide some kind of fragmentation algorithm, as is the case for IP.

Ethernet Subscriber Access

Now that we have a view of the different building blocks that can be used to build a layer-2 network of any size, let's have a look at the ways in which network operators can generate revenue from such networks. It is important to note that "Ethernet Subscriber Access" relies on a very different set of requirements depending on whether the subscribers are residential users or business users. The high-level view of these network services in this section will allow us to assess different deployment models in the next section.

Residential Services

A typical residential Ethernet service would at the very least offer high-speed Internet access. To generate more revenue for the service provider, this service can be combined with next-generation services such as voice-over-IP and video (broadcast and on-demand), resulting in the so-called "triple play."

Classical IP—IP straight on top of Ethernet—is not suited for subscriber access for several reasons. Ethernet relies on MAC address uniqueness for correct delivery of datagrams, and this is not something one can absolutely count on in a noncontrolled environment; there is equipment that allows user reconfiguration of the MAC address, and there may be equipment that has a nonunique MAC address as it leaves the factory. Therefore, another kind of user identification must be added to the datagram to allow correct forwarding. This is where ATM has come in handy in the past.

On point-to-point IP links, such as dial-up Internet connections, the Point-to-Point Protocol (PPP) is used for session setup and identification. On bit-synchronous or byte-synchronous links, PPP is usually encapsulated in an HDLC-like manner.[12] A special version of PPP exists for use on Ethernet networks. This PPP-over-Ethernet (RFC 2516) protocol encapsulates and identifies individual IP sessions and provides the necessary authentication and control mechanisms.

Today, PPPoE is widely used for the autoconfiguration of IP services over DSL [43]. It was introduced in DSL access networks as a solution to scalability and service provisioning issues that arose with the rapid increase in popularity of residential DSL services [24], which were provisioned up till then by means of ATM virtual circuits. PPPoE (over

---

[12] See RFCs 1661 and 1662 for details on PPP and HDLC-like encapsulation.

ATM) was preferred over PPPoA, because it allows different end stations at the subscriber premises to be connected at the same time, using simple and cheap Ethernet LAN equipment (ATM LAN equipment never saw any significant adoption). In an EFM-based access network—where there are no ATM virtual circuits to guide the subscriber traffic to the service provider's network edge—PPPoE could be used to automatically configure subscriber hosts and tunnel subscriber traffic through the aggregation network.

Of course, other protocols than PPPoE exist to configure a host for an IP session, such as such as the Dynamic Host Configuration Protocol (DHCP, see RFC 2131). DHCP does not encapsulate the subscriber's IP frames, however, so another form of flow marking is still required if the traffic is to be transported through the provider's layer-2 network. The use of VLAN bridges and/or provider bridges for this purpose was discussed previously. Another limitation of DHCP is that it does not support authentication; it can, however, be combined with the port-based access control mechanisms specified in IEEE Std. 802.1X—which we discussed in Chapter 9—to get a complete solution.

A different access solution is to completely separate the layer-2 network of the subscriber from the layer-2 network of the provider by placing a layer-3 gateway or router in-between. The limited address space of the Internet Protocol v4 (IPv4)—only 32 bits and running out fast in a world where more and more devices are being networked—brings about the need for a network address translation scheme at the home router or at the access multiplexer. The introduction of IPv6, which has an expanded address space and a number of "flow-oriented" features, seems very promising for the access and aggregation network.

Business Services

The Metro Ethernet Forum quotes ease of use, cost-effectiveness, and flexibility as the benefits of Ethernet services [36]—as compared to frame relay services, TDM, or private leased lines. The most naked business Ethernet service that could be deployed, is an "Ethernet private LAN" or "Ethernet Virtual Connection" (EVC), in which a subscriber uses different Ethernet[13] access links connected to a switched provider network, which to the subscriber, behaves exactly like a pri-

---

[13] According to [36] these links should comply to IEEE Std. 802.3-2002. This must be understood to include the amendments to IEEE Std. 802.3-2002 specifying 10 Gigabit Ethernet and EFM. It is expected that future MEF white papers will reference the latest revision of IEEE Std. 802.3.

vate LAN. The provider network itself may contain portions that are based on provider-bridged Ethernet, MPLS, ATM, or other protocols, which do not require participation from the subscriber.

Just like an actual Ethernet LAN, the EVC must avoid sending frames back to the user-network interface (UNI) from which they originated and keep the frame unchanged (i.e., the EVC should offer MA-UNITDATA.indication instances with parameters identical to those of the MA-UNITDATA.request instance that caused it). In MEF speak, Metro Ethernet services are called "E-Line" if two UNIs are involved, or "E-LAN" if multiple UNIs are involved. Note that several service instances may be multiplexed over the same physical interface; in this case, VLAN tags are used to label the different services.

E-Line. An E-Line connects two subscriber sites to each other through a switched provider network. If best-effort service is insufficient, the service level agreement between the subscriber and the service provider may specify performance parameters such as bitrate (guaranteed bitrate and peak bitrate), frame loss, and delay/jitter performance. If minimal delay and jitter, and zero frame loss are requirements, for example, to mimic the characteristics of TDM private leased line, the physical interfaces between the subscriber and the provider network must not be used for any other service instances.

E-LAN. An E-LAN connects two or more subscriber sites to each other through a switched provider network, thus emulating a bridged network. If best-effort service is insufficient, the service level agreement between the subscriber and the service provider may specify performance parameters such as bitrate (guaranteed bitrate and peak bitrate), frame loss, and delay/jitter performance.

The E-LAN service is more easily extensible than a set of point-to-point services between the same N end points; the addition of one end point requires provisioning of only additional UNIs, instead of having to provision N additional bidirectional point-to-point links.

Deployment Models

In shared and bridged Ethernet networks, a variety of issues can cause frames to be visible to end stations other than the one for which they are intended.

- Shared networks (e.g., 10BASE2, 10BASE5) and semi-shared networks (e.g., EPON) rely on an architecture in which frames are physically visible to all attached end stations.
- Bridges may "flood" frames if the destination address was not previously assigned to a specific bridge port through management or through the learning mechanism.
- Certain higher-layer protocols, such as ARP,[14] use broadcast messages to obtain or distribute certain information, which are normally forwarded to all ports of a bridge.

In Chapter 9, we discussed encryption as a way of keeping subscriber traffic from being read by an unintended recipient, but for traffic engineering and reliability reasons, we better keep data from reaching unintended recipients in the first place. Likewise, encryption does not protect the network from Denial-of-Service attacks, in which a malicious user disrupts the normal operation of the network by sending false configuration messages or by overloading the network with "legitimate" traffic. To deal with these issues, a high level of traffic segregation between different users (subscribers and service providers) is required. VLANs can offer this segregation.

There are unlimited ways to combine C-VLANs and S-VLANs into an adequately segregated subscriber access network, some of which make more sense than others. To keep the number of options manageable for users and for system vendors, several specialized configurations have been proposed. Two such solutions are worth discussing in more detail: the *VLAN cross-connect* and the *intelligent bridge*.

The solutions discussed here are taken from [18]; other system vendors may propose slightly different configurations. There are no standard deployment models for Ethernet access networks yet, although efforts are underway in the DSL Forum (Working Text 101, which deals with "Migration to Ethernet Based DSL Aggregation," was started in 2004).

VLAN Cross-Connect

The VLAN cross-connect configuration mimics the connection-oriented nature of an ATM-based access network by assigning a different VLAN to every subscriber port on the access multiplexer. Inside the access

---

[14] This is the Address Resolution Protocol used by IP. For more on TCP/IP and how it interacts with Ethernet, see [40].

multiplexer, each subscriber lives in a VLAN consisting of only her own port (carrying untagged traffic) and a network uplink port (carrying tagged traffic). Forwarding decisions at the access multiplexer become trivially simple, as all incoming subscriber traffic goes to the uplink, and all incoming network traffic goes to a single subscriber port, as indicated by the VLAN tag.

Assuming that the subscriber has no use for VLAN tags, all traffic in the first mile can indeed be presumed untagged. The first access multiplexer (remote unit, DSLAM, or other access node) can then be a relatively simple Customer Bridge, using C-VLANs to set up the cross-connect. The next access multiplexer (aggregating DSLAM or switch) must add an S-VLAN tag, to keep the cross-connect model scaleable, or use higher-layer forwarding mechanisms.

The VLAN cross-connect model will do nicely for residential subscribers. Business users, on the other hand, may want to use a certain set of C-VIDs to manage their own traffic and expect these C-VIDs to be respected by the provider network. In that case, the access multiplexer needs to use the S-VLAN space to set up the cross-connect or preassign certain sets of C-VLANs to business users.

Hybrid models in which certain S-VIDs designate a single (business) subscriber and others group several (residential) subscribers distinguished by C-VIDs may be configured in a single access network. This could be the case in a single access node or in a hierarchical network consisting of remote units (assigning C-VIDs) and access concentrators (assigning S-VIDs).

The disadvantages of the cross-connect model are its lack of inherent support for multicast (although this can be worked around), the artificial and configuration-intensive solution for mixing residential users with business users, and the scalability limitations.

Intelligent Bridge

The VLAN cross-connect is admittedly a bit ATM-ish. That is not necessarily a bad thing; in a network that brings together parties that are not necessarily friendly to each other or the service provider, a connection-oriented architecture—or an emulation thereof—offers a number of security guarantees that will limit the damage that a malicious user can bring about.

The Ethernet paradigm is, however, connectionless: Frames are forwarded individually on the basis of dynamically obtained topology information. To securely deploy a bridged network, certain precautions need to be taken. These are collectively referred to as "intelligent bridging" or "residential bridging."

The basic philosophy behind intelligent bridging is that malicious or ill-inspired actions performed by one subscriber should not affect the network performance experienced by any other subscriber. To this end, a number of user-to-user communication mechanisms are blocked.

- Broadcast frames are intercepted to avoid broadcast storms and related denial-of-service attacks. Valid broadcast frames, such as ARP, DHCP, or PPPoE protocol messages are either forwarded to the appropriate network port, or acted on locally by the access multiplexer. Other broadcast frames are discarded. Broadcast frames from a subscriber line are never forwarded to other subscriber lines. Broadcast frames from the network pertaining to a particular subscriber are never forwarded to other subscribers.
- Subscriber MAC addresses are hidden from other subscribers. Business subscribers requiring layer-2 connectivity between sites connected to the network through different subscriber lines should be protected from the rest of the network by means of a dedicated S-VLAN. Within that S-VLAN, normal ARP operation is allowed. Between subscribers, ARP messages are intercepted, allowing the access multiplexer to force all inter-subscriber traffic through a router.
- The number of MAC addresses that any one subscriber can use is limited. Occurrence of duplicate MAC addresses between subscribers is detected, so port-hopping and the related denial-of-service attacks can be avoided.

In the intelligent bridging paradigm, access multiplexers should not bother with C-VLANs; these can safely be used by (business) subscribers for their own purposes. The access multiplexer should add an S-VLAN tag to the traffic going out on the network port, to allow for efficient traffic engineering in the aggregation network. The assignment of S-VLAN IDs can be static (per access multiplexer, or per set of subscriber ports on the access multiplexer), protocol-dependent, or even destination MAC address dependent. This last case can be used to deploy a multi-service network in which the subscriber obtains the MAC addresses of the service routers through the usual ARP process, and the access multiplexer segregates the traffic going to these service routers on the basis of the destination MAC address present in the subscriber frames.

The special measures required for intelligent bridging can already be implemented in some commercially available bridge components. Intelligent bridging therefore has a better claim to being "the Ethernet way" of deploying an access network.

Conclusions: The New Ethernet

In the 30 years of its existence, Ethernet has evolved from a technology to connect computers to a 10-Mbps coax-cable "Ether," to a family of standards spanning speeds from 1 Mbps to 10 Gbps and features as diverse as management, flow control, OAM, authentication, and security. These elements can be combined to design an economical, secure, high-speed subscriber access network. An example lay-out is shown in Figure 10.4.

The migration from ATM-based access networks to Ethernet-based access networks is inspired by the lower cost associated with Ethernet equipment—thanks to Ethernet's mass market economics. However, the enhancements to traditional Ethernet that are currently being

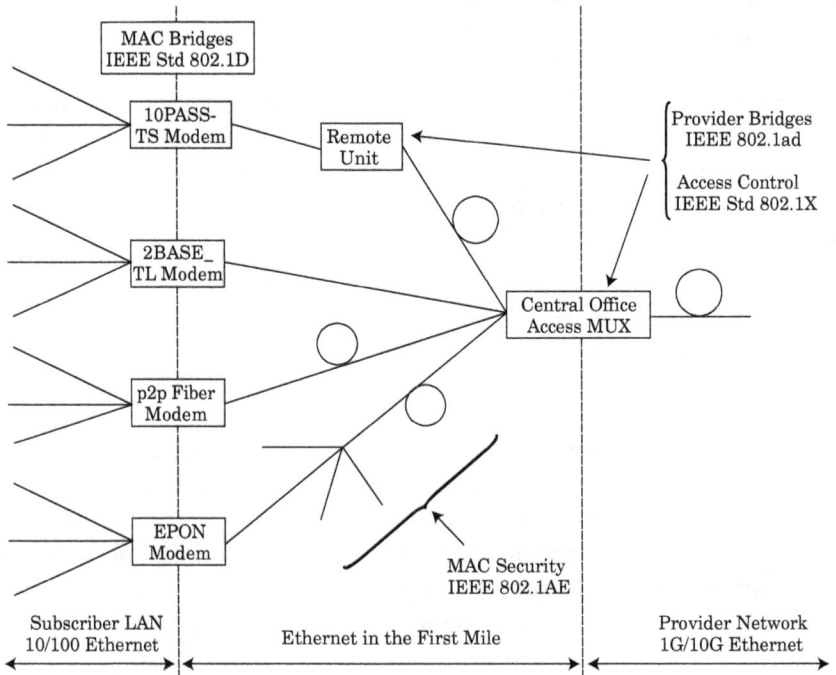

Figure 10.4   Ethernet-based access network.

standardized to make Ethernet "carrier-grade" will introduce additional complexity and cost....

In a way, this effort aims at making Ethernet more like ATM. But the lesson learned from ATM is that complexity must not be overdone. The incremental approach of Ethernet should allow it to stay on top of the mass-market wave while adding functions that allow network operators to deploy value-added services.

A similar statement can be made about the relationship of the "new Ethernet" to IP. Many of the layer-2 functions described in Chapters 8, 9, and 10 have a layer-3 counterpart in the IP protocol suite, as shown in Table 10.1. This seems to imply a certain duplication of effort between the IEEE LMSC and the IETF. There are, however, reasons for adding these functions to Ethernet.

TABLE 10.1   Functional Mapping of Ethernet Protocol Suite to IP Protocol Suite

Ethernet Protocol (Draft/Standard)	Internet Protocol (RFC)
EFM PME Aggregation (802.3ah)	Multilink Protocol
EFM OAM (802.3ah)	SNMP
Port Authentication (802.1X)	PANA
Service VLANs (802.1ad)	MPLS
Provider Backbone Bridges	L2TP
MAC Security (802.3AE)	IPSec
Connectivity Fault Management (802.1ag)	ping/traceroute/ICMP

For any network of considerable size, many functions beyond mere data forwarding are *required* if any value is to be extracted from the network. Although Ethernet and IP are like a marriage made in heaven, Ethernet is occasionally deployed without IP in the higher layers. In these cases, the value-added functions must either be replicated in the layer-3 protocol of choice or relocated in the layer-2 infrastructure.

The advantage of enriching the layer-2 infrastructure with the features we have discussed is that the network will be manageable regardless of the layer-3 protocol running over it. This is important when the business model of the network operator consists of leasing pure layer-2 VPNs to customers, which means that the operator has no control whatsoever over the layer-3 data that run over the layer-2 network.

Also, the segments between consecutive IP hops may consist of several intermediate Ethernet hops. These nodes do not have IP suite protocols at their disposal, but need nevertheless be configured, managed, authenticated, and the like.

The desired result of the enhanced functions in Ethernet is that relatively cheap switching equipment will be capable of offering services that used to require more expensive routing equipment. This statement, although often heard, may seem to be flawed. The very reason why routing equipment tends to be more expensive for a similar data forwarding capacity, is the presence of these ancillary functions and protocols. Surely, adding the same functions to any other device would increase the cost of that device to the same extent?

The difference lies in the mass-market economics of Ethernet. The high volume of Ethernet devices being produced allows for a short product lifecycle and a gradual introduction of features, even if many features are not used. As a result, the product cost can be kept under control.[15] All these factors make that it is very likely that Ethernet technology will become more and more prevalent in subscriber access and aggregation networks in the years to come.

---

[15] This implies that off-the-shelf Ethernet devices will often include features that are used by only a small portion of the market, but paid for by the entire market.

A

EFM Objectives

These are the objectives of the IEEE 802.3ah "Ethernet in the First Mile" Task Force, as approved by the IEEE 802.3 Working Group on July 11, 2002. (Numbering added by the author for referencing from within this book.)

1. Support subscriber access network topologies:
   - Point-to-multipoint on optical fiber
   - Point-to-point on optical fiber
   - Point-to-point on copper

2. Provide a family of physical layer specifications:
   - 1000BASE-LX extended temperature range optics
   - 1000BASE-X ≥10 km over single SM fiber
   - 100BASE-X ≥10 km over SM fiber
   - PHY for PON, ≥10 km, 1000 Mbps, single SM fiber, ≥1:16
   - PHY for PON, ≥20 km, 1000 Mbps, single SM fiber, ≥1:16
   - PHY for single pair nonloaded voice grade copper, distance ≥750 m and speed ≥10 Mbps full-duplex
   - PHY for single pair nonloaded voice grade copper, distance ≥2700 m and speed ≥2 Mbps full-duplex

3. Support far-end OAM for subscriber access networks:
   - Remote failure indication
   - Remote loopback
   - Link monitoring

4. The point-to-point copper PHY shall recognize spectrum management restrictions imposed by operation in public access networks, including:
   - Recommendations from NRIC-V (USA)
   - ANSI T1.417-2001 (for frequencies up to 1.1MHz)
   - Frequency plans approved by ITU-T SG15/Q4, T1E1.4, and ETSI/TM6

5. Include an optional specification for combined operation on multiple copper pairs

6. Optical EFM PHYs to have a BER better than or equal to 10^{-12} at the PHY service interface

The EFM Modem

Introduction

In the world of telecommunications, "modem" is the traditional name for an apparatus that *mo*dulates and *dem*odulates a signal to transfer digital information over a public network infrastructure such as the Public Switched Telephone Network (PSTN). For many years, the word modem was synonymous with "voiceband modem," a modem designed for transmission of data within the narrow frequency band that is normally used to carry voice conversations; in recent years, the word has commonly been used to designate broadband access devices (e.g., cable modems, DSL modems).

There are, however, a number of important differences between voiceband modems and broadband access devices.

- Voiceband modems really only care about getting the bits onto the telephone wire. Broadband access devices, on the other hand, will often provide some form of data encapsulation, because the terminals will generate data in discrete packets, whereas the medium expects a steady flow of bits or bytes. The encapsulation is necessary to fill up the gaps between the packets (e.g., idle cells in ATM or flag bytes in HDLC) and to indicate to the other end where packets begin and end, a task performed by end-terminal software in the case of voice-band modems (PPP, SLIP). These functions were discussed in more detail in Chapter 3 for the point-to-point copper PHYs.
- The owner of a voiceband modem can use it to directly call any other compatible modem; the analog signal of the modem is switched through the PSTN from one modem to the other, such that an end-

to-end connection is established at the physical layer. Even when the data call is terminated at a service provider's modem bank, the modems on both ends of the link are technically identical. Broadband access devices imply an asymmetric architecture: The end user's modem connects to an access multiplexer (DSLAM, head-end, OLT), which terminates the physical layer and carries the data over to an aggregation network that connects to the service provider's point-of-presence (POP).

- Broadband access devices, as a general rule, cannot be used to connect to another broadband access device at the physical layer; communication between end users will always be transported over an intermediate layer-2 (or above) network. The need for this dedicated equipment at the network side of the subscriber line puts the owner of the equipment—the network access provider—in the subscriber access value chain. This fact has an impact on business models, as was discussed in Chapter 10.

In Ethernet, there is no established use for the word "modem," as the standards tend to focus on the functional decomposition of a device rather than its implementation as a real-world product (hence definitions of functions such as the Physical Layer Entity [PHY], Medium Attachment Unit [MAU], repeater, and bridge). In this appendix, I will refer to an EFM-based broadband access device as an Ethernet in the First Mile (EFM) modem, regardless of its position in the OSI model. It is less common to speak of the subscriber line termination at the operator side as a modem, so this use of the word will be avoided. The operator side of the EFM link was discussed in Chapter 10.

As explained in Chapter 1, the EFM standard is about new PHYs. A PHY all by itself is of surprisingly little use to an end user. The higher layers (typically TCP/IP) can only access the services of a PHY by means of a MAC sublayer, shown in Figure 1.3, which is responsible for generating the MAC frame, including the source and destination address, a type/length field, the payload data, and a 32-bit CRC. To people who are new to Ethernet, it is often unclear where the Media Access Control (MAC) sublayer actually resides in a communication system. The following examples clarify this.

- In a PC, the MAC and the PHY are traditionally placed on a Network Interface Card (NIC); modern machines may have a NIC integrated on the motherboard. The higher layers of the protocol stack, which are implemented in software as part of the PC's operating system, can access the network hardware via an internal bus. A PC with a NIC can be interconnected directly with another PC having a compatible NIC, in which case the MACs of both PCs communicate directly with each other. The PHYs may be interconnected through a shared medium (coax), through point-to-point links to a repeater, or through a direct point-to-point link (using either a crossover cable or an integrated crossover function in one of the NICs).

- An Ethernet switch[1] is a device with multiple Ethernet ports, each of which consists of a PHY (or an interface to which a PHY can be connected) and a MAC connected to a common MAC Relay entity. The MACs in a switch do not transmit frames with their own unique source address; instead, they transmit the MAC frames that they get from the MAC Relay entity with the original transmitter's MAC source address. The MAC Relay does not have the same delay constraints as a hub, as it may perform buffering and priority queuing. Each of the ports has its own buffer, so each PHY can independently be configured for full-duplex operation, without the risk of collisions that would arise on a shared LAN; a switch is therefore said to separate collision domains.

- A router is a layer-3 gateway. Like a switch, a router is a multi-port device in which each port has its own MAC. The difference is that a router attaches a layer-3 (IP) entity connected to logic that can route packets between the ports based on their layer-3 addresses.

Media Converter

If you think of Ethernet as the framed data stream that flows out of the interface on your computer (presumably 100BASE-TX), you may be tempted to devise an "EFM modem" as a device that takes in that data stream and outputs the same data onto one of the EFM media in real-time. Such a device, consisting of a traditional Ethernet PHY and an EFM PHY back-to-back, could be MAC-less; it would be functionally equivalent to a 2-port repeater or a "media converter."

Straightforward as it may seem, this approach has severe drawbacks. First, both ports would have to run at the exact same bitrate, because there is no practical way to control the rate on the traditional Ethernet segment from the EFM PHY. Second, in case of the optical EFM PHYs, there would be no place for the EFM-specific changes to

---

[1] This is a particular implementation of a "MAC bridge," as specified in IEEE Standard 802.1D. See Chapter 10 for more information on bridges.

the operation of the MAC Control, MAC, and Reconciliation Sublayers, so not only would such an implementation not have all functionality, but it would also be noncompliant to the standard. The media converter solution is only feasible for the EFM point-to-point optical PHYs, when combined with a traditional Fast Ethernet or Gigabit Ethernet PHY.

The media converter example illustrates the fact that the correct way of thinking of Ethernet is as a *service provided by the MAC*, which relies on the PHY to access a physical medium.

Network Interface Card

A first architecture that is compatible with this view is an EFM NIC. This is a combination of an EFM PHY and an Ethernet MAC on a single PC extension board, which offers a MAC Service interface to the processor via an internal bus (e.g., the PCI bus). Such a system is architecturally identical to existing non-EFM NICs and internal DSL modems.

The architecture is not substantially different if the internal bus is replaced by an external bus (e.g., USB) or a point-to-point link. The external modem still offers its MAC service interface to a single host.

The drawback of the NIC or external modem solution is that the EFM link is only made available to one end station; that is, it cannot be shared.

EFM Bridge

Architecture

In a more general solution, the EFM PHY is combined with a MAC by connecting it to a Media Independent Interface (MII) of a bridge. The end result is very similar to a local area network (LAN) switch. In Customer Premises Equipment (CPE), the EFM port would serve as the wide area network (WAN) port or "uplink," while the other port(s) of the switch could be equipped with typical LAN interfaces such as 10/100BASE-T, WiFi, etc. All users of the LAN can then access the EFM segment through the switch.

In a Central Office (CO) design, things are the other way round. The switching fabric will connect a single high-speed WAN port (e.g., Gigabit Ethernet or 10 Gigabit Ethernet) to a high number of EFM ports, which serve the different subscriber lines.

The basic premise of transparent bridging is that frames received on one port are forwarded to all other ports. This allows hosts on different LANs connected to the same bridge to communicate with each other as if they were on a single LAN. Two important exceptions to the forwarding rule are the learning process and the spanning tree protocol. The learning process lets bridges remember for a limited time (typically 5 minutes) on which port frames from a particular MAC address were received; subsequent frames addressed to this MAC address will be forwarded to that port only. The spanning tree protocol blocks certain ports to avoid loops in the topology of the bridged LAN.

An example of a bridge-based EFM modem is shown schematically in Figure B.1. The EFM Copper PHY can be replaced by any other EFM PHY type, assuming that the capacity of the LAN interfaces is in an adequate relation to the capacity of the uplink. The processor is added to allow management of the bridge and optionally of the PHYs. Additional components such as external memory and power supplies are not shown.

Existing broadband access devices are often constructed in a similar way, with additional adaptation functions between the Ethernet bridge and the WAN PHY (in the case of ADSL modems, this would include an ATM adaptation layer).

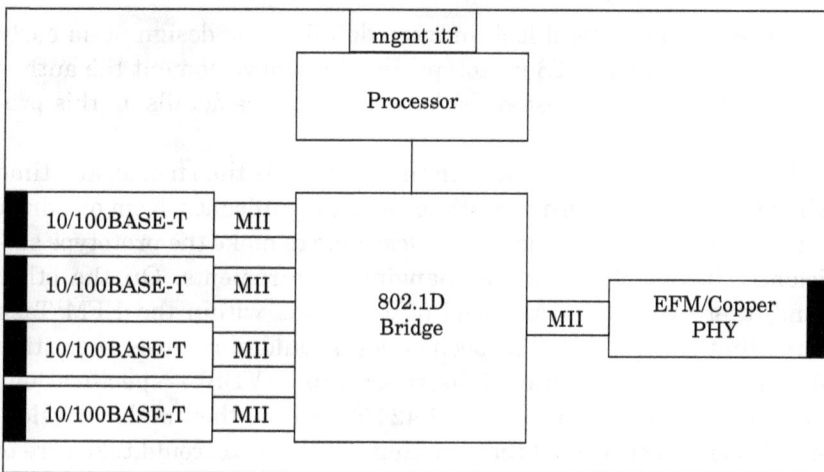

Figure B.1 Example of an "EFM copper modem," consisting of four LAN ports and a single EFM copper WAN port.

Note that there may be several EFM interfaces attached to the bridge. In that case, the bridge should allow aggregation of the EFM links into a single logical interface through the appropriate aggregation protocols (PME aggregation for EFM copper; link aggregation for Fast and Gigabit Ethernet).

The two-port bridge can be considered as a special case of an Ethernet switch. Its function is similar to that of a media converter in that it connects two different media segments together to form a hybrid end-to-end link, yet it is more intelligent than a media converter because it contains two MACs back-to-back. It can thus perform rate matching; terminate the Operations, Administration, and Maintenance (OAM) protocol; be managed, etc.

However, the bridging functionality it contains is heavily overdesigned for this simple two-port topology; functions like MAC address learning, forwarding, and multicasting really become trivial in a two-port bridge, yet their implementation remains required for compliance with the Bridged LAN standards. It is expected that a future amendment to IEEE Std. 802.1D will specify a simplified architecture for two-port bridges, thus defining a "smart media converter" that will provide a very simple, standard-compliant way to introduce EFM terminations into existing switched Ethernet networks.

Case Study: 10PASS-TS Prototype

As a case study, we will look in more detail at the design of an early bridge-based 10PASS-TS prototype, in whose development the author participated. The interested reader can find more details on this prototype in [9].

The prototype was developed in the 2001–2002 time frame, at a time when the EFM standard was still evolving significantly from one draft to the next. For this reason, it was desirable to make the prototype sufficiently flexible to adapt to changing requirements. On the other hand, there was a certain level of consensus within the EFM Task Force that the 10PASS-TS specification should stay close to existing Very-High-Speed Digital Subscriber Line (VDSL) specifications (specifically, either linecode of T1.424/Trial-Use, then under development), so it was expected that existing VDSL silicon could take care of most of the Physical Medium Dependent (PMD) and Physical Medium Attachment (PMA) functions.

Field-Programmable Gate Array (FPGA) technology was selected to rapidly and flexibly implement the digital logic required for EFM copper (PCS and TC sublayer). The FPGA was positioned between an Alcatel DMT-VDSL chip set and a commercially available LAN-switch component. The Alcatel DMT-VDSL chip set could be configured to operate either as a VTU-O or as VTU-R.

At first, it seems sufficient to implement an MII, a PCS, a TC sublayer, and an appropriate interface toward the VDSL chipset to have a complete 10PASS-TS modem. This could indeed be a gate-optimized version of a 10PASS-TS prototype. In reality, we chose the more complex approach shown in Figure B.2.

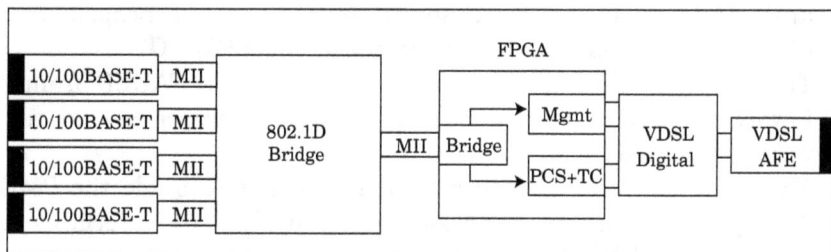

Figure B.2   10PASS-TS prototype architecture.

To fulfill its role as a research prototype and a testbed for different modem configurations, the prototype (or at least the PHY portion) had to be *manageable*. The standard way to make a PHY manageable is to provide an MDIO interface toward the manager. However, the Clause 45 MDIO registers for 10PASS-TS were not sufficiently evolved at the time. Moreover, the need for frequent software and firmware downloads made the possible use of MDIO a bit cumbersome.

For practical reasons, we preferred running a lightweight proprietary management protocol in-band over a UDP/IP/Ethernet stack. This implied we needed a UDP/IP/Ethernet protocol stack in the FPGA and a small block of logic capable of recognizing the management frames in the data stream and diverting them to the internal management entity. As this forwarding process is based on the frames' destination MAC address, the logical block is functionally equivalent to a bridge, though not an IEEE Std. 802.1D-compliant one (learning is not supported).

As a result, the prototype could be managed from any PC attached to the bridge by means of a small Java application, not unlike the way end users can manage (some functions of) their broadband access devices today. The prototype could also be managed from a PC attached to a modem on the other side of the 10PASS-TS link, when configured to allow this.

The interface between the FPGA and the VDSL chipset consists of a data part and a management part. The data interface into the digital VDSL bypasses the digital VDSL's built-in ATM-TC functions, connecting directly into the byte-synchronous α(β)-interface. Data are pushed to and pulled from the FPGA by the VDSL chipset.

Advantages of this architecture are:

- It provided the necessary flexibility to keep up with changes in the draft standard, such as the change from HDLC-based encapsulation (PTM-TC) to 64/65-octet based encapsulation (EFM-TC).
- The internal bridge could generate IEEE 802.3x PAUSE frames toward the MII, to remedy the fact that CRS deference was not correctly supported by the external bridge.
- There were sufficient resources left in the FPGA to add the PME Aggregation Function, using additional physical α(β)-interfaces to connect to additional VDSL chip sets.
- There are sufficient resources left in the FPGA to experiment with alternative encapsulations, such as PPP in HDLC-like encapsulation and IP in 64/65-octet encapsulation, which use the available bandwidth more efficiently.
- There are sufficient resources left in the FPGA to add and test pre-standard or nonstandard enhancements, such as dual latency (which requires a multiplexer/demultiplexer and two instantiations of the PCS and TC sublayer).
- The Operations, Administration, and Maintenance (OAM) protocol could be implemented in the FPGA.

Disadvantages are:

- FPGA technology is economically suboptimal for large volumes.
- The management protocol is inefficient; it clutters the PHY, reducing its density and augmenting its cost.

As a conclusion, the approach described above was sensible for a research prototype, but a commercial development would benefit from integrating the PCS and TC sublayer either into the digital PMA/PMD chip or in a dedicated ASIC.

The internal bridge can probably be left out in an optimized architecture. It is expected that future LAN switches will all support carrier deference properly (hopefully, even when operating in full-duplex mode). Management can be centralized on the board, interfacing with the PHY through an MDIO or similar interface. It remains to be seen if OAM appears on the feature list of LAN switches.

EFM Router

A further enhancement of the EFM modem concept would be the addition of layer-3 awareness; the EFM modem then becomes a router. Assuming the Internet Protocol is used as the layer-3 protocol, forwarding of frames between the LAN ports and the EFM port is then performed on the basis of the destination IP address of the frame.

There are no limits to the higher-layer functions that a more advanced CPE modem could offer in addition to mere routing:

Firewalling

Firewalling ensures a secure separation between the LAN and the public network; this prevents unauthorized traffic (intrusion attempts, viruses) from entering the local network. More complex traffic screening can be used to filter out spam and suspicious files.

PPPoE Termination

PPPoE encapsulation is often used to tunnel subscribers' IP traffic through the access and aggregation network. When PPPoE is terminated in the router, the WAN port can be treated by the rest of the local network as any other IP port accessible through the router. PPPoE and/or PPPoA termination is already a common feature of broadband access routers.

Dynamic Host Configuration Protocol (DHCP)

DHCP is a protocol that lets a router dynamically assign IP addresses from a pool to attached hosts. It is often used in conjunction with NAT.

Network Address Translation (NAT)

NAT allows different end stations to share a single public IP address. The NAT device "translates" the source IP address of outgoing packets (which may be an address from a special address range that has local significance only) into the public IP address assigned to the WAN port of the router (this can be a statically assigned IP address, or one provided to the router through PPPoE or DHCP). To distinguish requests from different local hosts, the source port number of the outgoing packet is also modified and a mapping between local IP addresses and port numbers is kept in a database.

Remote Configuration

As broadband access devices grow more complex, they require more skills to configure them properly. A subscriber who has configured his own access device may not have the means to distinguish a local configuration error from a network failure, which could lead to unnecessary (and costly) help desk interventions or house calls. It can, therefore, be beneficial for the network operator to directly configure the subscriber's equipment through dedicated protocols.

Conclusion

The modem architectures and features described above line up well with those of existing cable modems, ADSL modems, and VDSL modems. The only differences result from the fact that the EFM transceivers are architecturally pure Ethernet PHYs that connect *directly* to the switching fabric or to the MAC. The management and OAM framework that exists for Ethernet can add extra value to EFM-based architectures. The development time for EFM-based modems can, therefore, be relatively short.

References

Standards, Recommendations, and RFCs

Alliance for Telecommunications Industry Solutions

Committee T1 standards can be purchased from ATIS at its website, www.atis.org. Note that the DSL standardization activities of Committee T1 have recently been taken over by the newly formed Committee NIPP, the former Subcommittee T1E1.

ATIS Committee T1E1.4 (Digital Subscriber Line Access). *T1.417-2003—Spectrum Management for Loop Transmission Systems.* ATIS T1E1.4, 2003.
ATIS Committee T1E1.4 (Digital Subscriber Line Access). *T1.424-2004—Very-high-speed Digital Subscriber Line (VDSL) Metallic Interface (DMT based).* ATIS T1E1.4, 2004.

American National Standards Institute

X3.230-1994—Fibre Channel Physical and Signaling Interface. ANSI, 1994.
X3.263-1995—Fibre Distributed Data Interface (FDDI)—Token Ring Twisted Pair Physical Layer Medium Dependent (TP-PMD). ANSI, 1995.

Institute of Electrical and Electronics Engineers

IEEE 802 standards older than 6 months can be obtained free of charge at standards.ieee.org/getieee802.

IEEE Std. 802-2001—IEEE Standards for Local and Metropolitan Area Networks: Overview and Architecture. IEEE, 2001.
IEEE Std. 802.1AB-2005—Standard for Local and Metropolitan Networks: Station and Media Access Control—Connectivity Discovery. IEEE, 2005.
IEEE Draft P802.1AC/D0 –DRAFT Standard for Media Access Control (MAC) Service definition. IEEE, 2003.
IEEE Draft P802.1ad/D4.0 –DRAFT Amendment to IEEE Std. 802.1Q—Provider Bridges. IEEE, 2005.
IEEE Draft P802.1AE/D3.0–DRAFT Standard for Local and Metropolitan Networks: Media Access Control (MAC) Security. IEEE, 2005.
IEEE Draft P802.1af/D0.1–DRAFT Amendment to IEEE Std. 802.1X—Authenticated Key Agreement for Media Access Control (MAC) Security. IEEE, 2004.
IEEE Draft P802.1ag/D2.0—DRAFT Amendment to IEEE Std. 802.1Q—Connectivity Fault Management. IEEE, 2005.

IEEE Draft P802.1ah/D0.2—DRAFT Amendment to IEEE Std. 802.1Q—Provider Backbone Bridges. IEEE, 2005.

IEEE Std. 802.1D-2003—IEEE Standards for Local and Metropolitan Area Networks: Media Access Control (MAC) Bridges. IEEE, 2003.

IEEE Std. 802.1Q-2003—IEEE Standards for Local and Metropolitan Area Networks: Virtual Bridged Local Area Networks. IEEE, 2003.

IEEE Std. 802.2-1998 [ISO/IEC 8802-2]—Information technology—Telecommunications and information exchange between systems—Local and metropolitan area networks—Specific requirements—Part 2: Logical link control. IEEE, 1998.

IEEE Std. 802.3-2002—Carrier sense multiple access with collision detection (CSMA/CD) access method and physical layer specifications. IEEE, 2002.

IEEE Std. 802.3ae-2002—Carrier sense multiple access with collision detection (CSMA/CD) access method and physical layer specifications—Amendment: Media Access Control (MAC) Parameters, Physical Layers, and Management Parameters for 10 Gb/s Operation. IEEE, 2002.

IEEE Std. 802.3ah-2004—Carrier sense multiple access with collision detection (CSMA/CD) access method and physical layer specifications—Amendment: Media Access Control Parameters, Physical Layers, and Management Parameters for Subscriber Access Networks. IEEE, 2004.

IEEE Standards Association. *IEEE SA Bylaws.*

IEEE Standards Association. *IEEE SA Operating Rules.*

LAN/MAN Standards Committee. *IEEE PROJECT 802—LAN MAN STANDARDS COMMITTEE (LMSC)—POLICIES AND PROCEDURES (Formerly known as OPERATING RULES OF IEEE PROJECT 802 LMSC,* 2003.

Internet Engineering Task Force

IETF RFCs are available from the Internet Engineering Task Force website at www.ietf.org/rfc.html.

Postel J, ed. *RFC 793—Transmission Control Protocol.* Internet Engineering Task Force/DARPA, September 1981.

Simpson W, ed. *RFC 1661—The Point-to-Point Protocol (PPP).* Internet Engineering Task Force, July 1994.

Simpson W, ed. *RFC 1662—PPP in HDLC-like Framing.* Internet Engineering Task Force, July 1994.

Sklower K, Lloyd B, McGregor G, et. al. RFC 1990—*The PPP Multilink Protocol (MP).* Internet Engineering Task Force, August 1996.

Rigney C, Rubens A, Simpson W, Willens S. *RFC 2138—Remote Authentication Dial In User Service (RADIUS).* Internet Engineering Task Force, April 1997.

Kent S, Atkinson R. *RFC 2401—Security Architecture for the Internet Protocol.* Internet Engineering Task Force, November 1998.

Mamakos L, Lidl K, Evarts J, et. al. *RFC 2516—A Method for Transmitting PPP Over Ethernet (PPPoE).* Internet Engineering Task Force, February 1999.

Harrington D, Presuhn R, Wijnen B. *RFC 2571—An Architecture for Describing SNMP Management Frameworks.* Internet Engineering Task Force, April 1999.

Bormann C. *RFC 2686—The Multi-Class Extension to Multi-Link PPP.* Internet Engineering Task Force, September 1999.

Bormann C. *RFC 2687—PPP in a Real-time Oriented HDLC-like Framing.* Internet Engineering Task Force, September 1999.

McCloghrie K, Kastenholz F. *RFC 2863—The Interfaces Group MIB.* Internet Engineering Task Force, June 2000.

Flick J. *RFC 3635—Definitions of Managed Objects for the Ethernet-like Interface Types.* Internet Engineering Task Force, September 2003.

Flick J. *RFC 3636—Definitions of Managed Objects for IEEE 802.3 Medium Attachment Units (MAUs).* Internet Engineering Task Force, September 2003.

International Organization for Standardization

ISO standards can be purchased from the ISO Store at www.iso.org.

ISO/IEC 3309 Information Technology Telecommunications and Information Exchange Between Systems—High-level data link control (HDLC) procedures—Frame Structure. ISO/IEC, 1993.

ISO/IEC 7498-1: 1994, Information processing systems—Open Systems Interconnection—Basic Reference Model—Part 1: The Basic Model. ISO/IEC, 1994.

ISO/IEC 15802-1: 1995, Information technology—Telecommunications and information exchange between systems—Local and metropolitan area networks—Common specifications—Part 1: Medium Access Control (MAC) service definition. ISO/IEC, 1995.

ISO/IEC 60793-2—Optical fibres—Part 2: Product specifications. ISO/IEC, 1992.

International Telecommunication Union

ITU-T Recommendations can be purchased from ITU-T at its website www.itu.int/publications/index.html.

ITU-T Recommendation G.975—Optical fibre submarine cable systems—Forward error correction for submarine systems. ITU-T, 2000.

ITU-T Recommendation G.984.3—Gigabit-capable Passive Optical Networks (G-PON): Transmission Convergence Layer Specification. ITU-T, 2003.

ITU-T Recommendation G.991.2—Single-pair high-speed digital subscriber line (SHDSL) transceivers. ITU-T, 2001.

Revised ITU-T Recommendation G.991.2—Single-pair high-speed digital subscriber line (SHDSL) transceivers. ITU-T, 2004.

ITU-T Recommendation G.993.1—Draft VDSL Recommendation Foundation Document. ITU-T, 2001.

ITU-T Recommendation G.994.1—Handshake procedures for digital subscriber line (DSL) transceivers. ITU-T, 2004.

Books, Articles, Conference Papers, and Other Publications

1. *The Ethernet, Version 2.0* (the "Blue Book"). Digital Equipment Corporation, Intel, Xerox, November 1982.
2. *Reduced MII interface.* AMD Inc., Broadcom Corp., National Semiconductor Corp, and Texas Instruments Inc., 1997.
3. *Serial-MII Specification.* Cisco Systems, 2000.
4. Beck M. *Specification of Public Access Networks (Part I).* Available on-line at www.ieee802.org/3/efm/public/nov01/beck_2_1101.pdf, November 2001.
5. Beck M. *Ethernet over VDSL: DMT Concepts and Advantages.* Available on-line at www.ieee802.org/3/efm/public/mar02/beck_2_0302.pdf, March 2002.
6. Beck M, Borghs E, et. al. "Ethernet Transport over Digital Subscriber Lines." In: *IEEE Benelux Chapter on Communications and Vehicular Technology—9th Symposium on Communications and Vehicular Technology in the Benelux,* pages 35–39, October 2002.
7. Beck M, Mihanta A. *Dual Latency in xDSL.* Available on-line at www.ieee802.org/3/efm/public/mar02/beck_1_0302.pdf, March 2002.
8. Bingham JA. *ADSL, VDSL, and Multicarrier Modulation.* John Wiley & Sons, 2000.
9. Borghs E, Jacobs J, et. al. "Prototyping Ethernet in the First Mile over Point-to-Point Copper." *Proceedings of the Thirteenth IEEE International Workshop on Rapid System Prototyping.* IEEE, July 2002.

10. Bostoen T, et. al. *Dynamic Spectrum Management in Practice*, Alcatel Technology White Paper. Available on-line at www.alcatel.com, July 2004.
11. Cavendish D. "Operation, Administration, and Maintenance of Ethernet Services in Wide Area Networks," *IEEE Communications Magazine*, pages 72–79, March 2004.
12. Chiruvolu G, Ge A, Krogfoss B. *Encapsulation schemes to extend Ethernet to Metropolitan Area Networks: A comprehensive analysis of popular and evolving encapsulation schemes for Metro-Ethernet*, Alcatel Technology White Paper, available on-line at www.alcatel.com, February 2004.
13. Chiruvolu G, Ge A, Elie-Dit-Cosaque D, Ali M, Rouyer J. "Issues and approaches on extending Ethernet beyond LANs." *IEEE Communications Magazine*, pages 80–86, March 2004.
14. Christensen CM. *The Innovator's Dilemma*. Harper Business, 2000.
15. Cioffi J. *Space: The Final Frontier (in DSL)*. IEEE Workshop on Emerging Access & In House Transmission Technologies (Antwerp, Belgium), April 2001.
16. Cunningham D, Dawe P. *Review of the 10Gigabit Ethernet Link Model* (ONIDS 2002 White Paper). Agilent Technologies, 2002.
17. Davidson S. "Testing LANs Optically to the Gigabit Ethernet Standard." *IEEE Spectrum*, pages 86–90, September 1997.
18. Davy D, et. al. *Multi-Service Ethernet Broadband Access Solutions*, Alcatel Technology White Paper. Available on-line at www.alcatel.com, September 2004.
19. Faulkner D. *Where Next for Fibre Access?* IEEE Workshop on Emerging Access & In House Transmission Technologies (Antwerp, Belgium), April 2001.
20. Finn N. *Standardizing Ethernet Provider Services, in Extending IEEE 802 Technologies to Residential and Public Networking: Internetworking and Security Aspects*. ESuP Telecommunicació Universitat Pompeu Fabra, May 2004.
21. Frazier HM Jr. "Media Independent Interface: Concepts and Guidelines." In: *WESCON/'95. Conference record. 'Microelectronics Communications Technology Producing Quality Products Mobile and Portable Power Emerging Technologies,'* pages 348–353, November 1995.
22. Frazier H. *Why Here, Why Now?* Available on-line at www.ieee802.org/3/efm/public/nov00/frazier_1_1100.pdf, November 2000.
23. Frazier H. *Draft PAR and 5 Criteria*. Available on-line at www.ieee802.org/3/efm/public/mar01/par_1_0301.pdf, March 2001.
24. Ginsburg D, Hattar M. *Implementing IP Services at the Network Edge*, Addison-Wesley, 2002.
25. Helmers SA. *Data Communications: A Beginner's Guide to Concepts and Technology*. Prentice Hall, 1989.
26. Kimpe M, et. al. *Long Reach PHY Proposal: SHDSL—Decision Time*. Available on-line at www.ieee802.org/3/efm/public/jan03/copper/kimpe_copper_1_0103.pdf, January 2003.
27. Kramer G. *On Configuring Logical Links in EPON*. Available on-line at wwwcsif.cs.ucdavis.edu/~kramer/papers/llid_config.pdf.
28. Kramer G, Pesavento G. "Ethernet Passive Optical Network (EPON): Building a Next-Generation Optical Access Network." *IEEE Communications Magazine*, pages 66–73, February 2002.
29. Lucero S. *Ethernet in the First Mile (EFM): Like A Bridge Over Troubled Water*. In-Stat/MDR, November 2002.
30. Lucero S. *Ethernet in the First Mile (EFM): Provisioning Broadband on the Cheap*. In-Stat/MDR, August 2003.
31. Marasco AA. *IPR and Standards: Legal Considerations*. Available on-line, ANSI, November 13 2001.
32. Marris A. *MAC-PHY Rate Adaptation Proposal*. Available on-line at www.ieee802.org/3/efm/public/jan02/marris_1_0102.pdf, January 2002.
33. De Prycker M. *Asynchronous Transfer Mode: Solution for Broadband ISDN*. Prentice Hall, 1995.
34. Rezvani B, et. al. *EFM and 2.5 kft—10 Mbps, symmetric 998-symm and 998 spectrum study*. available on-line at www.ieee802.org/3/efm/public/nov01/rezvani_3_1101.pdf, November 2001.

35. Robert HM III, et. al. *Robert's Rules of Order Newly Revised*, Perseus Publishing, 2000.
36. Santitoro R. *Metro Ethernet Services—A Technical Overview*, Metro Ethernet Forum, 2003.
37. Seifert R. *The Switch Book: The Complete Guide to LAN Switching Technology*, John Wiley & Sons, Inc., 2000.
38. Sklar B. *Digital Communications, Fundamentals and Applications*, Prentice Hall International Editors, 1988.
39. Starr T, Cioffi JM, Silverman P. *Understanding Digital Subscriber Line Technology*. Prentice Hall, 1999.
40. Stevens WR. *TCP/IP Illustrated, Volume 1: The Protocols*. Addison-Wesley Professional Computing Series.
41. Suzuki H, Finn N, et. al. *EPON P2P Emulation and Downstream BroadCast Baseline Proposal*. Available on-line at www.ieee802.org/3/efm/public/mar02/suzuki_1_0302.pdf, March 2002.
42. Technical Report TR-003—*Framing and Encapsulation Standards for ADSL: Packet Mode*, ADSL Forum, June 1997.
43. DSL Forum Technical Report TR-044—*Auto-Configuration for Basic Internet (IP-based) Services*. DSL Forum, December 2001.
44. Urricariet C. *XFP: The Ultimate 10Gb/s Transceiver*. Available on-line at www.kogent.co.jp/product/finisar/pdf/xfp_article_rev_e_200212.pdf, December 2002.
45. Vaughan-Nichols SJ. "Achieving wireless broadband with WiMax." *Computer*, 37, June 2004.
46. Wilson SG. *Digital Modulation and Coding*, Prentice Hall, 1995.
47. Wright GR, Stevens WR. *TCP/IP Illustrated, Volume 2: The Implementation*, Addison-Wesley Professional Computing Series, 1995.

Index

ABOUT THE AUTHOR

Michael Beck is a research engineer at Alcatel Research and Innovation. As Editor of the EFM Copper Sub Task Force, he had a major role in the creation of the Ethernet in the First Mile standard. Mr. Beck holds a Master's Degree in Physical Engineering from Ghent University and has written numerous papers on EFM. He resides in Antwerp, Belgium.

www.ingramcontent.com/pod-product-compliance
Lightning Source LLC
Chambersburg PA
CBHW050456190326
41458CB00005B/1301